KW-361-870

The Politicisation of Islam

The Politicisation of Islam

A Case Study of Tunisia

Mohamed Elhachmi Hamdi

Westview Press

A Member of the Perseus Books Group

State, Culture, and Society in Arab North Africa

Published in 1998 in the United States of America by Westview Press, 5500 Central Avenue, Boulder, Colorado 80301-2877, and in the United Kingdom by Westview Press, 12 Hid's Copse Road, Cumnor Hill, Oxford OX2 9JJ

Library of Congress Cataloging-in-Publication Data
Hamdi, Mohamed Elhachmi
The politicisation of Islam : a case study of Tunisia / by Mohamed Elhachmi Hamdi
p. cm. — (State, culture, and society in Arab North Africa)
Includes bibliographical references (p.) and index.
ISBN 0-8133-3458-6 (hc) —ISBN 0-8133-3888-3 (pb)
1. Harakat al-Nahdah (Tunisia). 2. Islam and politics—Tunisia.
3. Islam and state—Tunisia. 4. Tunisia—Politics and government.
5. Islam—Tunisia—History—20th Century I. Title. II. Series
BP64.T7H33 1998
322.4'2'09611—dc21 98-52574
CIP

The paper used in this publication meets the requirements of the American National Standard for Permanence of Paper for Printed Library Materials Z39.48-1984.

PERSEUS POD ON DEMAND

10 9 8 7 6 5 4 3 2 1

To the memory of my father, Yousef

Contents

Acknowledgments

First and foremost, I would like to extend my thanks to both Dr. Katharine Zebiri and Dr. Michael Brett of the School of Oriental and African Studies (University of London) who were a great source of advice and guidance when I was preparing this study during my Ph.D. programme at the School. They were both patient and generous with their time when discussing my views, reviewing my progress and helping me to consolidate my arguments in a clear and solid manner.

I am also indebted to my wife Zoubida Guemadi for the enormous assistance and support she offered me during the six years it took to prepare the material for this book and for ensuring that I completed the task, even when other concerns and engagements threatened to divert me from academic life.

I should also like to thank Mrs. Zaynab Alawiye who was helpful in giving her time and effort to review the text prior to publication, despite having other commitments to her family and business.

System of Translation and Transliteration

All translations from the Arabic and French are mine, unless otherwise indicated.

The standard system of transliteration has been used for Arabic terms, names and sources throughout the text, with minor modifications: the initial ḥamza has been omitted, as has the final 'h' of the tā' marbūta.

All Arabic names of persons originating from the countries of the Maghreb have been transliterated according to the French system; therefore shīn becomes 'ch' (as in the name Ghannouchi), and the qāf is rendered as a 'k.' In instances where these individuals have authored sources in Arabic, their names have been transliterated in the notes according to the Arabic system of transliteration; thus Ghannouchi becomes 'al-Ghannūshī'. If these same individuals have also written works in French or English, they will be referred to in these cases in accordance with the French system.

Certain Arabic words and names of personalities that are of common usage in English are not transliterated, for example 'Islam', 'Qur'an', and 'Gamal Abd al-Nasser'.

Introduction

The aim of this study is to discover and analyse the history and discourse of the Tunisian Islamic movement *al-Nahḍa* in relation to post-independence Tunisian history and the trends of thought within contemporary Islamic movements. As such, the study is both historical and analytical. It attempts to give an accurate reading of the emergence, rise and eclipse of *al-Nahḍa*, as well as a comprehensive analysis of its political, social and intellectual discourse.

The importance of the study comes from the fact that it may be considered the first academic research done in English on *al-Nahḍa*; up to now, the movement has received only scant treatment in English sources. The most elaborate work to date on this subject is François Burgat and William Dowell's *The Islamic Movement in North Africa.*[1] As the title suggests, however, the Tunisian case is only a part of a wider North African interest; Burgat originally published his findings in French in his book *L'Islamisme au Maghreb: la Voix du Sud,*[2] where he reviewed the history of the Islamic movement not only in Tunisia but also in Algeria, Libya and Morocco. None the less, his research into *al-Nahḍa* may be considered the most thorough thus far among French sources.

Even in Arabic, books published about the Tunisian Islamic movement are very rare. Of the few published, one is entitled *al-Nahḍa,*[3] and was written by the government official Abd Allah 'Imāmi, who studied the movement under the rubric of terrorist organisations in the Islamic world. A second is *al-Ittijāh al-Islāmī wa Bourguiba: Muḥākamat man li man?*[4] [The Islamic Trend and Bourguiba: Who is trying whom?], written by Walīd al-Mansūrī from within the ranks of *al-Nahḍa,* refuting the government's main charges at the major trials of 1987 and published in 1988. These two books formed part of the political war between the Tunisian régime and *al-Nahḍa* and do not constitute objective sources for academic research. I may add to this category my own book, entitled *Ashwāq al-Ḥurriyya,* which I based on a collection of *al-Nahḍa* documents and published in order to draw attention to the plight of Islamist detainees in Bourguiba's gaols.[5]

Although the Algerian Islamic Salvation Front (FIS) has attracted more attention in the Maghreb as a whole since 1990, it was the Tunisian Islamic movement that played the role of the champion of Islamism in the region during the eighties, gaining more fame and attracting more scrutiny than any other

Islamic organisation in the Maghreb. However, this scrutiny has come mostly from the media rather than from academics.

In this sense, carrying out academic research in English into *al-Nahḍa* may claim some originality from this fact alone, in that it may achieve the following: to encourage other researchers to give closer attention to an important part of the Maghreb, and to a movement that has given the impression of adapting to modernity and the changing times, succeeding in the late eighties in commanding strong sympathy among many democrats and liberals in the Arab world, and even in the West, which is supposedly seen as an ideological opponent of Islamism in general.

But there are also other factors that give this study its significance. Most important is that it aims to expose and analyse the efforts of the Tunisian Islamists in answering some of the major questions facing those who are engaged in contemporary Islamic thought, and specifically the role of Islam in modern public life. Probably the first question leading to this wider debate is how a political movement gains its Islamic identity or nature: is it simply because it calls itself Islamic? Is it because of its programmes or intentions, or because of the religious affiliations of its leaders and members?

As the conflict intensifies between Islamic movements and their governments, Tunisia itself being one of the most striking cases, questions are raised with regard to this claim to Islamicity, especially by opponents of Islamic movements, who accuse them of manipulating religion for political ends, and who call for Islam to be above partisan politics. However, this question is not only about politics, but also about essential matters of faith and understanding of Islam. This is particularly true when studying the view of Islamic movements, in this case *al-Nahḍa*, their societies and their religious mission. Being an "Islamic" movement poses a very important question: is the society itself Islamic? If it is not, should Islamists obey its rules? The answer to these questions may prove vital in reassessing the political role of religion in the Arab world, and the political and cultural divisions that arise from it.

Another angle from which to assess the religious nature of the political action and discourse of the Islamists is by studying their programmes. Since *al-Nahḍa* and most similar Islamic movements stress that their reforming mission is as comprehensive as Islam itself, this research offers an opportunity to assess the contribution of Tunisian Islamists to enhancing the compatibility between Islam and modernity, a question that has often been raised before by the major Muslim thinkers and reformists of the modern period, such as Jamāl al-Dīn al-Afghānī, Khayr al-Dīn al-Tūnisī, Muḥammad 'Abduh, 'Abd al-Raḥmān al-Kawākibī, Sayyid Quṭb and others.

This study attempts to explore and analyse the kind of "Islam" proclaimed by Tunisian Islamists, whether it reflects a classical understanding of religion, or a modern activist one which resembles that of contemporary Middle Eastern Islamic movements. The discourse of *al-Nahḍa* offers a useful opportunity in this context, especially because of the detailed project its leader elaborated on the definition of a modern Islamic government, one of the most important issues of heated debate between Islamists and their opponents in the Muslim world.

But if al-Afghānī and 'Abduh were essentially individuals who tried to answer the questions of modernity by intellectual discourse, *al-Nahḍa* and other political Islamic movements are different in that they are ready to organise their supporters and make a serious attempt to capture political power and bring about the sought-after Islamic state. That is why in addition to analysing the intellectual input of *al-Nahḍa*, this study tries to answer another equally significant question about the means employed by the Islamic movement. This provides an opportunity to study the activist aspect of the movement: how their political tactics reflect their religious credentials, and how the Western concept of a political party is used in a religious context, to the point of becoming an alternative social refuge for a misguided, or even a *jahilī* society. Other important issues in this regard are the openness or secrecy of the movement, the role of the internal organisation (*tazīm*), and the combination of civil and military plans to acquire political power. It is also in this context that the study tries to analyse the characteristics of the people who adhered to the movement's message and joined the *tazīm* in order to serve the aim of building an Islamic state.

To answer these questions, I have opted for a flexible methodology that recognises the multi-dimensional nature of the issue. Essentially, I have used a double-track method, by studying both the history and the discourse of the movement. I have attempted to offer what may be seen as an objective and up-to-date account of *al-Nahḍa's* history. This is a delicate task because of the conflicting nature of the available sources, which come mainly from either the movement itself or its opponents. Even secondary sources were usually supportive of one party or the other. To write the history of *al-Nahḍa* is to some extent to write the history of independent Tunisia, and that entails looking at various political, social and cultural factors, and analysing the positions of various players on the scene, including the government, the leftist opposition, and the trade unions, although without losing sight of the subject of this study. Even in its historical context, the research is concerned with analysing the history of ideas within the movement, how they emerged and how they have evolved and changed in line with changing political circumstances.

These ideas become the prime interest of the second half of the book. My method was to read the most important texts of the movement and analyse them in their Tunisian and pan-Islamic context. I especially concentrated on some key questions related to the meanings of "Islamicity" for *al-Nahḍa*, and on evaluating their contribution to reconciling the given teachings of religion to the changing times and needs of Muslims. This was achieved by examining the movement's discourse on three major topics: the comprehensiveness of Islam, the nature of the Islamic state and the Islamic identity of Tunisia.

Most of the written material examined in this book is that of Rachid al-Ghannouchi, the main founder of *al-Nahḍa* and its most prominent leader from the outset. He was both the political and religious leader of the movement, as well as an intellectual who expressed himself most openly about the main challenges facing the Islamists during the 1970s, 1980s and 1990s. There are a few documents officially declared to be in the name of the movement, and a few contributions by other less prominent members of the leadership; most of these have been examined in the study. The two most important names in the latter category are Abd al-Fattah Morou, who was secretary-general of the movement until 1991, and Abd al-Majid al-Najjar, who was more active on the intellectual side as a professor of theological studies. I have used some of their statements and writings where appropriate, although it should be mentioned that Morou resigned from the party in 1991, and al-Najjar announced in 1995 that he no longer had any links whatsoever with *al-Nahḍa*.

I have used mostly published material, and in a few cases unpublished documents that I have obtained during the last few years. In the historical section of the book, I was also able to verify facts from Ghannouchi himself, for he was living in London. I should mention here also that I have benefited from the fact that I was a member of the movement from 1978 to 1992, which covers most of the period under study in this book. I served as a leader of the student wing of the movement in the eighties and worked with the main Politbureau inside and outside the country. In the 1987 major trial of the movement, I was sentenced in absentia to 20 years in gaol. Earlier, in 1983, I spent more than 6 months in gaol with some prominent members of the leadership of the time. I was also one of two members of the movement who managed to speak directly with President Zein al-Abidine Ben Ali, the other being Ghannouchi. In this instance, I telephoned him directly from London in 1988 to seek assurances about the future of the movement and permission for myself and two other prominent leaders, Morou and Hammadi al-Jebali, to return to the country from exile.

I have tried to benefit from my long and close relationship with the movement in a positive way, that is to say in verifying certain historical facts and intellectual opinions and thereby offering what I hope will be seen as the most accurate reading of the movement's history up to the mid-1990s.

I have endeavoured to be as objective as possible; not only was this an academic obligation, but also a true desire to exercise an impartiality which I was not able to have when I was an activist for the movement. The mid-1990s gave me the chance to look back at a history that affected my life and the lives of thousands of my colleagues. My aim was not to defend it or to dissociate myself from it. My aim was to understand it, and I hope that this book will testify to that.

Notes

1. Austin: University of Texas Press, 1993.
2. Paris: Karthala, 1988.
3. Tunis: al-Dār al-Tunisiyya li-l-nashr, 1992.
4. Tunis: n.p., 1988.
5. Kuwait: Dār al-Qalam, 1989.

1

The Emergence of the Tunisian Islamic Movement

The history of the contemporary Islamic movement in independent Tunisia is basically that of the movement now known as *al-Nahḍa*. Founded initially as *al-Jamā'a al-Islāmiyya* (The Islamic Group), the movement changed its name in 1981 to *Ḥarakat al-Ittijāh al-Islāmī* (The Islamic Trend Movement), and in 1988 changed its name once more to *Ḥarakat al-Nahḍa* (The Renaissance Movement). Although some of its founders resigned in 1978 and formed a new group under the title *al-Islāmiyyūn al-Taqaddumiyyūn* (The Progressive Islamists), and a second group resigned in 1991 and tried to form a new political party, the core of the Tunisian Islamic movement remained loyal to *al-Nahḍa*, making it one of the main opposition parties in the country.[1] This chapter will analyse the economic, political, religious and cultural factors that were behind the emergence of the movement.

The first cell of *al-Jamā'a al-Islāmiyya* was set up in 1970. At that time, Tunisia was embarking on a new era of economic liberalisation after the failure of the socialist experiment led by Ahmad Ben Salah, a former minister who, under the supervision of Bourguiba, had been responsible for the country's economic sector.

Economic and Political Factors

Ben Salah was an influential leftist figure in the Tunisian General Workers' Union (UGTT), which had played an important role in the national struggle for independence, achieved formally on 20 March 1956. In this same year the UGTT organised its sixth congress and proposed a complete economic programme, suggesting a centrally planned and centrally oriented economy, managed by a council presiding over the ministries of reconstruction, agriculture, public works, post, telegram and telephone, finance, the central bank and the national economy.

According to Ben Salah, only the state could possibly preside over such a revolutionary economic plan, which aimed at dismantling the economic system created by colonial France and ending exploitation in all of its forms. This vision was presented as an alternative to what Ben Salah then described in one of his speeches to congress as "the liberal anarchy" that still reigned over Tunisia's social and economic plan.[2]

A centrally planned economy was the mood of the fifties and sixties in the majority of the newly independent countries of the Arab world, which found their main international ally in the Soviet Union in their fight against the colonial policies of Western European countries, particularly France and Great Britain. Although at the beginning of his term as prime minister in 1956 and as the first president of independent Tunisia in 1957, Bourguiba was not enthusiastic about the UGTT programme, he did not totally rule out the possibility of embracing Ben Salah's vision.

Indeed it took Bourguiba another four years to ensure that his position was not in danger from the ambitious Ben Salah, whom he had brought into the government in as early as 1957 to be minister of labour and health. In 1961 Bourguiba told the Tunisian people that the country needed "a new conception of national solidarity; a form of socialism."[3] He went on to explain the ideological basis for this important change of direction: "The Prophet's Companions in the first century of Islam were socialists before the term was invented. They regarded themselves as members of the same family. So let's return to the origins of Islam, as the employed is a brother to his employer." He added, however, "Although I've opted for socialism, I'm still opposed to the idea of class struggle."[4]

Ben Salah was thus promoted to the post of secretary of state for planning and to be a member of the ruling party's political bureau in 1961. Three years later the party changed its name from The Neo Destour to the Socialist Destour party, all to emphasise its new socialist strategy.

A National Council for Planning was also created. Under the supervision of Ben Salah, it devised an economic programme for the country's development, with a view to its completion within a ten-year period between 1962 and 1971. It also identified four essential and desired objectives, being the decolonisation of the economy, the promotion of the human being by ameliorating his financial situation, a reform of the country's traditional structure, and auto development, meaning the bringing about of a decrease in foreign debts and the inclusion of the entire country in central decisions.[5]

This programme was the brainchild of an influential group within the ruling Destour party. As Eva Bellin writes in "Tunisian Industrialists and the State," the programme's publication "marked the ascendance of the *dirigiste* wing of the political élite, an élite committed to setting Tunisia on a 'socialist' path to development."[6]

Apart from the above-mentioned goals, Tunisian socialism was also shaped around co-operatives that had been set up in the agrarian sector. Unfortunately this was a policy that did little more than to deprive farmers of the right to own their land, and in fact made them workers on land that had originally been their own but had been confiscated by the state. The official view concerning this policy was arrogant: it was argued that the farmers could not be given land because they were "illiterate, used to archaic methods of exploitation," besides which they were seen as "reactionaries and obstacles to progress."[7]

A new law was announced which made illegal every kind of land exploitation except that by the co-operative units of agricultural production. Thus, ironically, the peasants were the first to experience the injustices of the new socialist dream; people saw their properties confiscated by the state and found themselves working for the new co-operatives for 2 litres of cooking oil, a kilogram of sugar and a few kilograms of semolina each week.

In the first instance, the government did not have the courage to implement this new law in the more prosperous areas of the country, especially in the central and northern coastal regions, from which came the majority of the ruling political élite, including Bourguiba and Ben Salah. However, when it was finally decided at the beginning of 1969 to extend the system to all parts of the country, the hardship in the other regions in which the system had already been implemented was impossible to hide.

As has been the case with almost all socialist economies the world over, the abolition of private ownership to the benefit of the state brought with it corruption and inefficiency. Most of the money reserved for the programme in Tunisia was to be "lost" before reaching its final destination. What made things worse was that almost 40 percent of investments came from foreign loans, mainly from the United States. As for efficiency, below is a description of the alternative administration which was supposed to replace that of the illiterate peasants:

> For every co-operative a director and a technical director were appointed. They were rarely of peasant origins. They came from cities, and were unable to distinguish between a potato and a tomato. They did not have any idea about agronomy or climatology. On the other hand, those illiterate peasants knew their environment very well.[8]

The disastrous results of this corrupt and inefficient system were soon to lead to a public outcry. Not only had people been dispossessed of their properties and exploited by their government, but also national revenues had slumped and foreign debts had became a heavy burden on the entire country. Bellin has summarised the reasons for Ben Salah's ensuing fall as including "a clash of political personalities, the discontent of the rural bourgeoisie who were threatened by Ben Salah's plan to subject their land to co-operative control, the fiscal crisis faced by the state, and the bad luck of consecutive years of drought and poor harvests."[9]

Thus the people began to express their anger and resentment, and Bourguiba soon realised that the socialist experiment was threatening not only the country's stability and prosperity but also his own position as the undisputed leader of the nation. He therefore acted swiftly to save himself from blame. In September 1969, Ben Salah was stripped of all his ministerial posts except for education, which he lost the following November. In March 1970 he was arrested and accused of high treason, and in the May of that same year he was condemned to ten years' imprisonment and hard labour. Six months later Hedi Nouira, a liberal, was installed as the new prime minister.

In 1973 Ben Salah escaped from prison and fled to Switzerland, from where he began to issue statements condemning President Bourguiba for acting against the people in the interests of a privileged class. He also established himself as leader of the radical Popular Unity Movement (MUP), which had been declared illegal in Tunisia itself.[10]

During Ben Salah's radical decade in power there had been one very important incident on both the Arab and international scenes: the Six-Day War in 1967 between the Arab states and Israel, in which the Arabs had been ignominiously defeated. If the Ben Salah experience had proven to be fatal for the fortunes of socialism in Tunisia, the defeat in the war against Israel had had a similar effect on the ideology of Arab nationalism, spearheaded by President Gamal Abd al-Nasser of Egypt, and which had hitherto had wide appeal all over the Arab world, including Tunisia. The result of the war was clear and simple: pan-Arabism had failed the crucial test against Israel, and it no longer qualified to lead Arab efforts towards freedom, unity and progress.

These two ideological failures were to prove to be very important, at least for the few founding members of *al-Jamā'a al-Islāmiyya*, as we may see from the words of one of them, Dr Ehmida Enneifar, talking to the French scholar François Burgat:

> In Tunisia, there was first of all the departure of Ben Salah's team. The end of the experience was very brutal; the minister was imprisoned. But what was more important for a number of youths was that they had seen that the same government could be on the left and then suddenly change direction clearly to the right, with resolutely liberal economic options. Many of them were completely disoriented. The whole matter backfired on the Tunisian state because the ruling party had insisted firmly on a precise project in which to build a modern state; later on we came to realise that what took place was not only a change of government, but was also largely a proof of the absence of that project. Those who joined the Islamists' ranks were those who found nothing to be attached to, right or left; they were uprooted.[11]

The essence of this testimony is that the change from the co-operative socialism of Ben Salah to the economic liberalism of Hedi Nouira led to an ideological and identity crisis for many of those young Tunisians who could, at the time, afford to think and argue about politics. These were essentially students, university or secondary-school teachers, who were better off financially than the majority of the impoverished peasant population. Each student at that time would receive a monthly grant of 40 Tunisian Dinars (almost 80 US dollars), with which he could help his family; the salary of a teacher was even better. Although on campus there had been an active student movement led largely by Marxist groups, the failures of socialism had helped to open the way for the new, but deep-rooted influence of Islamic culture. According to Paul Balta, it was a "sizeable vacuum" in which Islamic ideology "could infiltrate without problem."[12]

Enneifer also mentions the 1967 defeat as a factor which may help to explain the emergence of the Tunisian Islamic movement:

> It was at that time that a number of intellectuals, including Ghannouchi and myself, began to meet. Very quickly, the question of religion was raised with insistence, because we could not find another way by which to proceed. Neither the Arab nor the Tunisian nationalists' ideas were able to offer an answer, nor even [those of] the West itself, which for a long time had appeared to us to possess absolute solutions.[13]

Referring to the 1968 student riots in French universities during the de Gaulle presidency, Enneifer comments that on his arrival in Paris, he thought he would find "a country in which questions were clearly defined or answered"; on the contrary, he argues, "What I found was the same disarray that we had. I realised then that it was not only a Tunisian or Arab, or Muslim problem. Even the West was passing through a dangerous era of general reflection about its way of life."[14]

In Tunisia at that time, as in almost all the Arab world, political and ideological trends were very much geared towards the search for a successful theory which would bring progress and prosperity, and which would put the country on an equal footing with the powerful nations of the West. The search for such a theory was yet another factor leading to the meeting of the first founding members of the Tunisian Islamic movement, particularly following Ben Salah's failures.

Writing in 1974, Ghannouchi asserted: "the fundamental question for our society and educational system is about the [sort of] model person we want: what his beliefs are, his philosophy, and how he [differentiates between] good and evil," adding: "If we were to ask those who plan our education about these matters, would they find an answer? Would they agree on an answer? The answer is most certainly 'no'."[15]

Both Enneifer and Ghannouchi seemed to be mainly concerned with the political and ideological dimensions of the failed socialist experiment. For a number of researchers and analysts, however, it is the economic factor that takes precedence in explaining the emergence and evolution of *al-Jamā'a al-Islāmiyya*; an analysis which may also be seen to apply to Nouira's liberal experiments. The basic argument behind the economic explanation is that the failure of both socialism and liberalism to deliver economic prosperity for a wide section of society led many of those affected to heed the Islamists' message. The Tunisian sociologist Elbaki Hermassi argues that personal sensitivity to what amounted to an economic threat made young Tunisians particularly susceptible to the charges of economic injustice, corruption, Western domination and exploitation put forward by Islamist groups against the government.[16] He also suggests: "the Islamist movement thrived on the ideological looseness of the post-independence state and has had the capacity to attract many of those left behind by economic growth."[17]

This argument is duplicated by Marion Boulby, another researcher on Tunisia's post-independence history, who writes: "the failure of Bourguibism to translate into employment of the educated youth cost the Tunisian régime dearly in terms of its own legitimacy. It is against this background that we note the development of Islamic revival."[18]

Susan Waltz notes also that of the three theses (political, economic and cultural) which may be seen to explain the emergence of the Tunisian Islamic Movement, "the economic one appears to be enjoying the greatest popularity."[19] She also refers to the contention that the current Islamist movement "is the most recent segment of an ongoing conflict between nationalist interests and the bazaar sector, who in protection of their traditional way of life seek to eliminate the new industrial ethics emphasising consumerism, women's emancipation, and promotion of culture and leisure."[20]

The Tunisian academic Dr 'Abd al-Majīd al-Sharfī has defended the same theory, albeit in different terms. He argues that the new Islamic movements recruited young people marginalised by unaccomplished modernisation, and that those who have responded to the Islamists' message are the unemployed youth and villagers migrating to the big cities, only to see their financial hopes dashed by failing economic policies. Proposing a solution to the "Islamist threat," al-Sharfī points clearly to an economic solution: "In the end we must be aware that neither speeches nor strong rational arguments will solve the problem of activist Islam. The solution depends on the continual breaking with an unjust international economic order."[21]

Religious and Cultural Factors

Whatever merits the economic and political explanation behind the emergence of *al-Jamā'a al-Islāmiyya* may have, Islamist leaders insist that their action was in fact a religious and cultural response to anti-religious and pro-Western policies. By way of explanation, Ghannouchi has illustrated Bourguiba's "attitude" towards Islam:

> In 1957, once in power, he prohibited the use of the *ḥijāb* and once uncovered a woman and tore her veil in public. Later, in 1981, a law was passed forbidding

> women employed in government offices or those entering universities and colleges to wear the *ḥijāb*. In 1957, he forbade polygamy.[22] These civil laws are still in force. In 1960, Bourguiba prohibited fasting in Ramaḍān, alleging that it was harmful to the country's economy.... In 1974 he stated that the Qur'an was self-contradictory and ridiculed the miracles of the Prophet Moses.[23]

In Ghannouchi's opinion, Bourguiba is little more than an "enemy" of Islam, but it is also interesting to see how other, more impartial researchers view him, and to examine how Bourguiba himself assesses his approach to Islam.

One observation which reflects the judgement of many Western and Arab writers about Bourguiba is that of Douglas K. Magnuson, a teacher at the Bourguiba Institute of Modern Languages in Tunis. He remarks:

> In the years after independence, Bourguiba embarked on a series of bold initiatives of religious reform affecting law, family life, education, and personal religious practice.
>
> His disassembling of the infrastructure of institutional Islam in Tunisia was so complete that social observers in the 1960s questioned whether Tunisia might have entered a post-Islamic or de-Islamicized age.[24]

The "bold initiatives" to which Magnuson refers are many. On 2 March 1956, Bourguiba abolished the traditional *ḥabūs* or *waqf* sector, which was founded on charity donations to mosques and religious schools, and which helped to maintain a level of financial independence for many Islamic scholars and institutions. The state confiscated all *ḥabūs* properties and made it illegal for anyone to offer new *ḥabūs*.

On 3 August in that same year, Islamic courts *(al-majālis al-shar'iyya)*, which used to preside over family law cases, were abolished to make way for new liberal laws that outlawed polygamy, among other things. This was backed by Bourguiba's campaign against the *ḥijāb*, which he described as a "sinister shroud that hides the face," even outlawing the veil in the classroom and going so far as to describe it as "an odious rag."[25]

In the first two years of independence Bourguiba closed the Zeitouna University, which had hitherto been a historical centre of Islamic learning, established in the second century of Islam and built before the famous al-Azhar mosque-university in Cairo and al-Qarawiyyīn in Fez. For Bourguiba, Zeitouna University was not at all needed for the building of the new state; on the contrary, it was seen as a dangerous obstacle both ideologically and politically.

On 5 February 1960, as Ghannouchi has remarked, Bourguiba appealed to the Tunisian people to stop observing the fast during the month of Ramaḍān because he claimed that it affected their capacity to work. Four years later he was seen on television drinking orange juice on the first day of Ramaḍān, encouraging people to break their fast for the sake of increased economic production. He told his people this was not contradictory to Islam, but that he was simply taking a "progressive" approach, which sacrificed fasting for the sake of a more important form of *jihād*, which was to develop the country's economy and prosperity.[26]

In order to replace Zeitouna University, Bourguiba set up a modern system of education similar to that of the French, and appointed an official mufti who had no real power whatsoever to criticise any official policies. However, according to the new constitution adopted in 1959, Islam was declared the official state religion and it was decreed that the president must be a Muslim.

It may be argued, however, that Bourguiba's real values came largely from sources other than Islam: "a child of the Enlightenment, educated in law and political science at the Sorbonne, Bourguiba took his political inspiration from Rousseau, Lamartine and Hugo. His goal was the recreation of Tunisia as a modern state according to the principles of the French Revolution."[27]

These were some of the facts which, as previously mentioned, led many people to question whether Tunisia might have entered a post-Islamic or de-Islamicised age. According to Habib Boularès, former minister of culture and information, Bourguiba wanted a modern country above all, "which implies engaging the country in the course of development, industrialisation and the widening of education," while adding that Bourguiba never hesitated to "follow the path of the West."[28] Boularès confirms also that Bourguiba envisaged the Tunisian state as "one freed of all shackles. This supposes the breaking of ties with old-fashioned habits, archaic traditions and structures of the past … and even certain forms of worship."[29]

Boularès argues that all this does not mean that Bourguiba was an "enemy" of Islam:

> Bourguiba learned the lessons of the experiment led by Kemal Attatürk thirty years earlier. He never fought against Islam itself.… Without claiming the title "Commander of the Faithful," Bourguiba insisted that the president of an Islamic state should be both a temporal and spiritual leader, by presiding personally over the ceremonies of the 27th night of Ramaḍān and the birthday of the Prophet.… He was always careful to justify each of his positions with a verse from the Qur'an, a saying from the Prophet's *sunna* or by an opinion from one of the past scholars and authorities.[30]

This is the same argument as that presented by the Tunisian writer Mohsen Toumi in an attempt to deny the notion that Bourguiba was an enemy of Islam, laying stress on the fact that Tunisia is a Muslim country and that its president must be a Muslim. "We have a habit of speaking about secularism when analysing the measures taken by Bourguiba after independence," Toumi argues. "It is a misinterpretation … because we ended up with the opposite result; the president became a kind of first *imām* playing the role of the country's spiritual leader."[31]

This final analysis of Bourguiba's attitude towards Islam comes from the perspective of those liberal and mainly leftist Tunisian writers who support Bourguiba's approach, and who feel it necessary to defend his achievements in the face of a new, radical wave of Islamism with which they can find almost no common ground. A more explicit testimony indicative of this trend is that of Ḥamma al-Hammami, leader of the extreme left Tunisian Communist Workers' party, who argues that Islam was in fact one of the pillars of the régime's ideology. He asserts that most of the social laws adopted by Bourguiba derive from *sharīʻa* laws, and that it was the government itself that paid for all religious activities in the country, because this served its own interests of defending the upper classes and exploiting the working classes.[32]

The Tunisian left was well aware that portraying Bourguiba as a religious leader would be useful in opposing the Islamists' call for a more religious political and social order in the country. Apart from the outright condemnation of critics such as Ghannouchi, and the sympathetic tendencies of liberals and secularists, Bourguiba's attitude to Islam is slightly more complex and difficult to define. He was not a communist preaching atheism, nor was he a disciple of Islamic reformers such as Jamāl al-Dīn al-Afghānī or Muḥammad ʻAbduh. In fact the model he tried to implement in Tunisia was Western—*à la française*—and he believed he could achieve this without breaking totally with Islam.

Naturally, his project could not follow the course of traditional or historical Islam, so he aimed to define a form of Islam that would be compatible with the needs of modern realities, and which could be achieved through *ijtihād*. Bourguiba supported this concept by asserting that Islam in fact "liberated the mind and recommended the rethinking of laws in order to adapt them to human evolution."[33]

Paul Balta notes that Bourguiba devoted many of his speeches to the necessity of reviving *ijtihād*. It is recorded that Bourguiba once said: "Islam has neither the Church nor monastic orders. Religious decisions are taken democratically by the community aided by scholars. Politicians have an imperative to bring about religious evolution suitable for the modern world."[34]

Ijtihād, an Islamic legal term, means "exerting oneself to form an opinion (*ẓann*) in a case (*qaḍiyya*) or on a rule (*ḥukm*) of law. This is done by applying analogy (*qiyās*) to the Qur'an and the *sunna*."[35] By the same token, the *mujtahid* is "one who by his own exertions forms his own opinion, being thus exactly opposed to the *muqallid*, 'imitator'.... "[36]

This definition in itself implies that any *mujtahid* is both an expert in Islamic rules and laws and a practising worshipper. Bourguiba may not have possessed these qualities, but insisted that he was eligible to make his own *ijtihād* as the Muslim leader of a Muslim country. Even in the case of his bold actions regarding fasting—one of the five pillars of Islam—he was adamant in rationalising this in terms of Islamic reference. In a speech given in Tunis on 18 February 1960 he argued that, similar to the cases of sickness or *jihād*, which constitute legal reasons for a Muslim to break his or her fast, the fight against under development and the struggle for economic prosperity is indeed also a form of *jihād* and thus should give grounds for legal abstention. He then went on to say: "In my capacity as head of a Muslim state, I too can speak in the name of Islam. The whole world knows what I did for Islam, at a time when some now-prudent professors showed [little more than] complacency towards the colonial régime."[37]

This was the kind of rhetoric Bourguiba used to promote his ideas, although they appear to be clearly contradictory to established Islamic notions, going so far as to present himself as a defender of Islam when his dispute with the Islamists became more dangerous and serious. For Ghannouchi and his friends, however, who began to form their Islamic movement in 1970, there was no acceptable excuse for the actions of the national leader. They were confident, Ghannouchi claims, that Bourguiba had been "active in fighting Islam and imposing on Tunisia all the values and ways of life of materialistic Western society."[38]

It was thus both a religious and cultural challenge which brought Ghannouchi and his friends together, or at least this was the fundamental motivation behind their actions. Hermassi acknowledges this factor when he says:

> As far as the emergence of the [Islamic] movement is concerned, there is a very obvious reason that is rarely taken into account or given the interest it deserves. It is the fact that, of all the Arab countries, Tunisia was unique in the public manner in which its modernist élites attacked institutional Islam and dismantled its basic institutions in the name of systematic social and cultural reform—the result was to dismantle the whole old cultural order. Besides that, this project was based on the resolution of the political leadership and was helped by the majority of the new graduates, and was accompanied by a very negative and contemptuous position towards traditional Islam. Was it for this reason and others that a few people gathered in 1970 around a magazine called *al-Ma'rifa* to exchange views about their alienation and that of their religion? Yes, there is no doubt about this matter.[39]

The Movement's First Founders

Among those who began to meet in 1970 to discuss their alienation and their religion were a number of young graduates from Arab or Tunisian universities, who shared the role of founding what is now the Tunisian Islamic movement. Among them were the three leading personalities who shaped the movement's form and ideology, being, in order of importance, Rachid al-Ghannouchi, Abdelfattah Mourou and Ehmida Enneifer.

Ghannouchi was born in 1941 to a poor family of ten children in al-Hamma, 30 kilometres west of Gabes, one of the main cities in south-east Tunisia. He says of his education:

> I completed my secondary education in the old Zeitouna *madrasa*, before it was closed down by the Tunisian government. I am of the generation of Zeitouna students during the early years of independence. I remember we used to feel like strangers in our own country. We had been educated as Muslims and as Arabs, while we could see the country being totally moulded in the French cultural identity. For us, the doors to any further education were closed since the university had been completely Westernised. At that time, those wanting to continue their studies in Arabic had to go to the Middle East. I was one of those who decided to complete their studies in the Middle East. I registered at the faculty of philosophy and letters in Damascus specialising in philosophy, and graduated four years later.[40]

As a young student, Ghannouchi was influenced by Arab nationalist ideas:

> When I began my university studies in 1964, the trend in the country was Arab nationalism, so I adhered to that for a period of time. Its content was scientific socialism, very close to Marxism. Consequently, during my first years of university I was a secularist. In my inner self, however, I did not cease to be a believer. I used to fast during the month of Ramaḍān but did not fulfil the prayers and other requirements of religion. I had always understood being an Arab and being a Muslim as inseparable realities, such as it is understood among our people in North Africa. In the Middle East, however, there are Christian Arabs and others belonging to various non-Islamic sects. There the concept of Arabism is very often in opposition to Islam.[41]

However, his affiliation to Arab nationalism did not last for long. He explains:

> When I came in touch with the other activists in the university who did not share the ideas of nationalism—the Islamists—I began a dialogue with them which progressively weakened the hold of Arab nationalism on my mind. After some time I realised that Arab nationalism was in opposition to Islam, while Arab sentiments and identity (in which I had been educated) and Islam were one and the same thing. At that time, I was a member of the Nasserite Nationalist party of Syria, but once I learned its true meaning I chose to abandon it and adopted Islam in its totality. Progressively, I felt more and more inclined to fight all those secular tendencies in each of their manifestations.[42]

Ghannouchi's statement shows clearly how the movement of ideas was taking effect on the sixties generation in the Arab world. Arab nationalism was the ideology of the time, Marxism was its social theory, while on the opposite side stood the Islamists, who were denounced by Nasser as reactionaries and enemies of his revolution. Meanwhile, inspired by the experience and ideas of the Syrian Muslim Brotherhood, Ghannouchi moved from the nationalist to the Islamist camp.

From exile in London in 1992, Ghannouchi spoke about what happened after his conversion to the ideas of the Islamists:

In June 1966, I finally renounced the Arab nationalist and secularist theories, and adopted comprehensive Islam. In 1968 I moved to Paris after getting my BA in Damascus, and there my relationship with Ehmida Enneifer became closer. I had already known him in Syria because he was a Nasserite Arab nationalist, but there had been few chances for us to meet and discuss political and ideological matters. In Paris, together with a few other friends, Enneifer and I found time to discuss a great number of issues. It was there that he distanced himself from Nasserism and became closer to Islamism.[43]

Ghannouchi had gone to Paris to study for a higher degree in philosophy, but was obliged to go back home only one year later owing to family circumstances. During his time in Paris, he had made contact with an Islamic group called *Jamāʿat al-Tablīgh*, and had become influenced by their methods. Once home, he embarked upon a career as a secondary-school teacher of philosophy. He explains:

I was fully converted to the views of the contemporary Islamic movement when I returned to my country. I settled in the capital and soon met Abdelfattah Mourou who used to attend the lessons of Sheikh Ahmed Ben Milad, a Zeitouna scholar. He was then still a student in Tunis, as was another in the first group we formed: Saleh Ben Abdallah. Fadhel Baldi also joined us; he was one of the students in the secondary school. Mourou was at that time strongly influenced by Sufi *madaniyya* law, but he quickly became more inclined to the activist aspect of modern Islamism. I kept in contact with Enneifer in Paris and exchanged letters with him. In one, he informed me that he had adopted Islamic views fully and pledged to join us when he came back. He returned in 1970 and became an influential member of the founding group.[44]

According to Ghannouchi, Mourou and Enneifer possess different qualities:

Mourou and Enneifer were the two most active members. Mourou has some unique interpretations of the Holy Qur'an and the sayings of the Prophet (peace be upon him), especially in matters of spiritual education. As an ex-Sufi he emphasised this dimension in the first generation of our movement, and received a very positive response. I myself learned a lot from him.

Enneifer represented the modern intellectual who used to read the French newspapers because of his fluent French. He has an intellectual ability to analyse various issues and was aware of what was taking place in the political arena. He was not a good orator, but he did have an organised way of thinking and arguing and an ability to organise his work, which had a considerable effect on our movement.[45]

Mourou had been a student in the faculty of law, from which he graduated as a judge before starting his own business as a lawyer. Enneifer returned to Tunis to become a teacher of Arabic language and literature in secondary schools. Both men came from the capital: Mourou from a modest family in the old quarter of Bab Souikha, and Enneifer from a well-known religious and old aristocratic family. Together with their friend Ghannouchi they began to develop the organisation of the new Islamic movement.

Embarking on a Religious Mission

In an interview with *Arabia* magazine, Ghannouchi talks of how the movement's work truly started in the 1970s, when a small group of young men formed in the capital. He says that at that time "it was difficult to find a young man praying, especially if he was from the so-called educated people. As for girls, it was almost impossible to see any dressed as a Muslim."[46]

He explains that the group had two levels of activity, promoting conferences and gatherings in secondary schools, and organising lessons on Islam in the mosques. Sometimes they would go out into the streets to call people to Islam, in the manner of *Jamā'at al-Tablīgh*.[47]

This group, originally founded in India in 1927 by Muḥammad Ilyās ibn Muḥammad Ismā'īl Qandahlāwī (1884–1943), served as an inspiration for the fledgling Tunisian Islamic movement. Qandahlāwī had feared that the minority Muslim community in India might lose its religious identity within the larger Hindu society, and had founded his group in order to revive Islamic beliefs among Muslims and to encourage them to observe Islamic teachings, to concentrate on acquiring Islamic knowledge and to worship Allah. The group did not favour participation in politics, believing that if individuals observed Islamic teachings, this would inevitably lead to an Islamic society.

As for its methods, *Jamā'at al-Tablīgh* insisted that its members must travel out of their provinces and even their countries to practise missionary activities, which they called *al-khurūj*. On arriving in a city or a village, they would invite people from their homes, the streets and the cafés to the mosque, where their senior leader would give a lecture on Islamic teachings. From India, the *Jamā'a* message spread to almost all the Islamic world and to Islamic communities in the West.

There were few indications at that time that Ghannouchi and his friends would have a realistic chance of success with their missionary activity, or *da'wa*, but to even their own astonishment, the movement began to attract more and more young members. They decided to join the official Association for the Safeguarding of the Holy Qur'an, initiated in 1970 by a group of traditional scholars who had succeeded in maintaining a good relationship with the government, and which was recognised and supervised by the department of religious affairs. For the Islamists, it was a good official cover for their early work,[48] for they could use the association's facilities to organise meetings and give lectures on Islamic affairs. The government, however, soon expressed its dissatisfaction with these new, enthusiastic members and ordered them out of the association. They then concentrated further on giving lectures in the various mosques around the country.

The movement at that time concerned itself mainly with strictly religious issues and the political interest was still very vague. As Enneifer argues, "We really did not know what we were aiming at. We generally disagreed with the government, but we did not have a well-defined plan of action."[49]

Calling people to observe the basic requirements of Islam, especially to attend the five daily prayers, and to be proud of both their Islamic history and identity were the main objectives of the Islamic movement at that time, which is why the methods of *Jamā'at al-Tablīgh* were deemed to be so appropriate. From 1970 to 1973 members of the movement travelled in groups to various villages around the country, "calling people from the streets, cafés and shops to listen to lessons on Islamic consciousness" and reminding them of their religious obligations.[50]

Magnuson has noted that it is also essentially the individual that the *al-Tablīgh* message and methods target, and that "their goal as a group is to create *ṣāliḥ* (righteous, virtuous, godly) individuals, as a means of arriving at a true Muslim society."[51] The influence of the *al-Tablīgh* group and its founders on its Tunisian followers was profound, even to the point of influencing how the Tunisian preachers dressed. In general, men's dress in Tunisia is a combination of Western and traditional styles, whereas *da'wa* men, by contrast, dressed in a manner that was modelled on the dress of Pakistanis and other Eastern Muslims. In fact, Tunisians often mistook *da'wa* men for foreigners.[52]

Such was the behaviour of the most committed followers of the methods of *Jamā'at al-Tablīgh*. Others, such as Ghannouchi and his colleagues, employed these methods but were not totally satisfied with them. Although he had participated in the activities of the *Jamā'at al-Tablīgh* group in Paris, where he had led the prayers for some time because he was an active preacher,[53] on his return from France Ghannouchi chose to immerse himself further in the literature of the Muslim Brotherhood.

Originally there had only been a few members of *Jamā'at al-Tablīgh* in Tunisia, and it had agreed to join hands with *al-Jamā'a al-Islāmiyya* because their main task in common was strictly religious. Ghannouchi, however, became increasingly preoccupied with the comprehensive view of Islam insisted upon by the Muslim Brotherhood. Co-operation between the two groups lasted until 1973, when police intervened to disperse a meeting of some ninety Islamists gathered in Sousse, preaching the *da'wa* in the *tablīghī* manner. Ghannouchi, Mourou and Enneifer were arrested and interrogated at the local police station, where they all insisted that they were *du'āt* (preachers); Mourou even claimed that they belonged to a 1,400-year-old party, referring to the Islamic religion itself.

Eventually the three were released and ordered to leave Sousse for their homes. In Ghannouchi's view, this represented a turning-point:

> We realised that the methods of *al-Tablīgh* may work in Pakistan or India, or even France where democratic traditions exist. But none of this was available in

> Tunisia. We decided to stop using *tablīghī* methods and to concentrate on the two main methods of *da'wa*: public lectures or printed publications on the one hand, and organising a number of regular, small and secret meetings in our homes to increase our knowledge of Islam and its teachings on the other.[54]

These meetings were to be the basic cells of the flourishing movement, and were adapted from the methods of the Muslim Brotherhood. In fact, from 1973 onwards, the influence of the Muslim Brotherhood would prove to be decisive on almost all levels.

By contrast, the influence of the *Tablīgh* group faded almost completely. Besides the security risks that its methods posed to the Tunisian Islamists, there was also the fact that the scope of its activities and its insistence on dealing with individuals failed to answer the wider questions that were prevalent in the Islamists' minds. The Islamic movement was opposed to the régime, even if it could not argue its case in detail, but the *tablīghī* message could not adapt to embrace wider social and political issues. Those in the movement soon realised that they had either to stick by this method or to adopt a new approach. They chose the latter.

It was during that same period (the early seventies) that Anwar Sadat of Egypt began to release the hundreds of Islamists imprisoned by his predecessor Nasser, which gave a new impetus to the Muslim Brotherhood's message and the propagation of its literature around the Arab world. Books by the Muslim Brotherhood were also made available in Tunis, where the first international exposition of books was held in 1973.

"It was precisely this movement which was to influence us," says Enneifer, "and which pushed us to be engaged more directly in political action as well as in setting up an underground organisation of sorts. We then started to form secret groups to provide it with an education, but always with a special spiritual dimension."[55] It was Enneifer who had the chance to meet some of the recently freed Muslim Brotherhood leaders in Mecca during the Ḥajj season of 1973, bringing back to his friends their advice, encouragement and good wishes.

Another important step was taken forward by the movement at around the same time. In 1972, Ghannouchi and his friends contacted Shaykh Abdelkader Ben Slama from the old Zeitouna establishment, and asked for his permission to relaunch *al-Ma'rifa* magazine. Although he had published only one issue in 1962, he still had the ownership of the magazine. Shaykh Ben Slama accepted the offer, and the founding group of Islamists began to put into action all their intelligence and creativity.

Their relaunch issue was swiftly confiscated by the authorities, because it criticised the abolition of the Islamic calendar from official publications and the use of the Western calendar to decide the beginning and end of Ramaḍān.[56] However, in *al-Ma'rifa* the group was able to voice its concerns and to call people to the true and comprehensive Islam in which it believed. The magazine's subject-matter was largely concerned with generalities, although Ghannouchi attempted some rather rudimentary analysis. He wrote in the tenth issue in 1973:

> Our country suffers from a number of fundamental problems, without their being thoroughly examined and analysed, such as the problem of morals, the problem of self-confidence, the problems of sex, individualism and the lack of readiness to offer help and sacrifice for others, cultural imitation, the problem of economic development and its relationship with morality, etc. These problems result from our historical, political and educational circumstances and from the effect of international circumstances on our own situation. It is necessary that we have our own criteria by which our circumstances should be judged. [By this] I mean a special culture and a special analysis of the universe, the human being and life.[57]

Ghannouchi then went on to condemn the Western values of the Tunisian education system and ended his article by describing an alternative culture with "Islam as its content and the Arab language as its framework."[58]

Such was the ideology of the movement up to 1978—these were young Islamists brought together largely by "general ideas, thoughts, and the trend of preaching about Islam," as Ghannouchi has described it, reflecting on his time in prison in 1982.[59] This was sufficient for that period in the movement's history, however; its members were busy spreading their message throughout the country, and as a result were receiving a positive response, as Salah al-Din al-Jourshi, another famous name from that era, explains:

> In the beginning we intended to address all sections of society.... But our experience showed that those who reacted most positively to our ideas and recommendations were youths. This is why there was up to a 70 percent majority of pupils and students in the movement. The other result of our experience was the widespread, positive response from the rural sectors. Whenever a teacher invited his pupils to observe Islam and the *da'wa* line, he found that those coming from rural areas were the most responsive. They then adhered to the *da'wa* and carried out its message.[60]

With the efforts and the dynamism of its predominantly youthful membership, the Islamist movement was almost guaranteed success, if only short-term. However, the potential for conflict and confrontations with its opponents was in evidence from the outset.

Notes

1. Each of these names will be used for the period in which it was used by the movement.

2. Extracts from Ben Salah's speeches and the UGTT programme are documented in Mohsen Toumi's *La Tunisie de Bourguiba à Ben Ali* (Paris: PUF, 1989), pp. 40–42.
3. Ibid., p. 55.
4. Ibid., p. 56.
5. Ibid., p.57.
6. Eva Bellin, "Tunisian Industrialists and the State," in *Tunisia: The Political Economy of Reform*, ed. I. William Zartman, SAIS African Studies Library (Boulder and London: Lynne Rienner Publishers, 1991), p. 48.
7. Toumi, *La Tunisie de Bourguiba à Ben Ali*, op. cit., p. 46.
8. Ibid., p. 62.
9. Bellin, in Zartman, ed., *Tunisia: The Political Economy of Reform*, op. cit., p. 63.
10. *The Middle East and North Africa 1980–1981*, no. 27 (1980), p. 743.
11. François Burgat, *L'islamisme au Maghreb: la voix du Sud (Tunisie, Algérie, Libye, Maroc)* (Paris: Karthala, 1988), pp. 204–5.
12. Paul Balta, *L'islam dans le monde: dossier établi et présenté par Paul Balta*, 2nd edn., Collection la mémoire du monde (Paris: Editions Le Monde, 1991), p. 144.
13. Burgat, *L'islamisme au Maghreb*, op. cit., p. 205.
14. Ibid.
15. Rāshid al-Ghannūshī, *Maqālāt* (Paris: Dār al-Karawān, 1984), p. 30.
16. Elbaki Hermassi, "The Islamicist Movement and November 7," in *Tunisia: The Political Economy of Reform*, op. cit., p. 193
17. Ibid.
18. Marion Boulby, "The Islamic Challenge: Tunisia since Independence," in *TWQ* no. 10 (April 1988), p. 599.
19. Susan Waltz, 'Islamist Appeal in Tunisia', in *al-Nahḍa* report, *The Movement of Islamic Tendency in Tunisia: The Facts* (Tunis: n.p., September 1987), p. 72. Waltz's article also appears in *MEJ*, no. 40 (1986), pp. 651–70.
20. Ibid., p. 72. See Von Sivers' 'Work, Leisure and Religion', in *Islam et politique au Maghreb* (Paris: CNRS, 1981).
21. 'Abd al-Majīd al-Sharfī, "al-Islāmiyyūn: a'dā' al-taḥdīth am ḍaḥāyāhu?," in *Al-Wiḥda*, no. 96 (September 1992), p. 47.
22. In fact the new personal status code was introduced in 1956.
23. Rāshid al-Ghannūshī, interview, *Arabia*, April 1985.
24. Douglas K. Magnuson, "Islamic Reform in Contemporary Tunisia," in Zartman, op. cit., pp. 169–70.
25. Bourguiba, at a speech given in Sfax, 5 December 1957; quoted in Boulby, "The Islamic Challenge," op. cit., p. 593.
26. See Abderrahim Lamchichi's *Islam et contestation au Maghreb* (Paris: L'Harmattan, 1989), pp. 89–90.
27. Boulby, "The Islamic Challenge," op. cit., p. 591.
28. Habib Boularès, *L'islam: la peur et l'espérance* (Paris: Editions J-C Lattès, 1983), pp. 197–8.
29. Ibid., p. 198.
30. Ibid., pp. 198–9.
31. Toumi, *La Tunisie de Bourguiba à Ben Ali*, op. cit., p. 116
32. Ḥamma al-Hammāmī, *Ḍidda al-ẓalāmiyya: fī al-radd 'alā al-ittijāh al-islāmī*, 2nd edn (Tunis: Dār al-nashr li al-maghrib al-'arabī, 1986), pp. 4–5

33. See Paul Balta and Claudine Rulleau, *Le Grand Maghreb: Des indépendances à l'an 2000* (Alger: Editions Laphomic, 1990), p. 61
34. Ibid.
35. 'Idjtihad', *Encyclopaedia of Islam*, 1971.
36. Ibid.
37. Habib Bourguiba, *Discours: Tome VII, Année 1959–60* (Tunis: Publications de Secretariat d'Etat à l'Information, 1976), pp. 142–4.
38. Rāshid al-Ghannūshī, interview, *Arabia*, April 1985.
39. Muḥammad 'Abd al-Bāqī al-Harmāsī, "al-Islām al-iḥtijājī fī Tūnis," in *al-Ḥarakāt al-islāmiyya al-mu'āṣira fī al-wa an al-'arabī* (Beirut: Markaz dirāsāt al-wiḥda al-'arabiyya, 1987), p. 250.
40. Rāshid al-Ghannūshī, interview, *Arabia*, April 1985.
41. Ibid.
42. Ibid.
43. Rāshid al-Ghannūshī, personal interview, 7 November 1992.
44. Ibid.
45. Ibid.
46. Rāshid al-Ghannūshī, interview, *Arabia*, April 1985
47. Ehmida Enneifar, quoted in Burgat's *L'islamisme au Maghreb*, op. cit., p. 205.
48. Al-Harmāsī, "al-Islām al-iḥtijājī fī Tūnis," op. cit., p. 259.
49. Ehmida Enneifar in Burgat, *L'islamisme au Maghreb*, op. cit., p. 206.
50. Magnuson, "Islamic Reform in Contemporary Tunisia," in Zartman, ed., op. cit., p. 170.
51. Ibid., p. 173.
52. Ibid., p. 174.
53. Rāshid al-Ghannūshī, personal interview, 15 November 1992.
54. Ibid.
55. Burgat, *L'islamisme au Maghreb*, op. cit., pp. 206–7.
56. Ibid., p. 206.
57. Al-Ghannūshī, *Maqālāt*, op. cit., pp. 11–14.
58. Ibid., p. 14.
59. Rāshid al-Ghannūshī, *Ḥarakat al-ittijāh al-islāmī fī Tūnis* (Kuwait: Dār al-Qalam, 1989), p.153.
60. Al-Harmāsī, "al-Islām al-iḥtijājī fī Tūnis," op. cit., p. 252.

2

The Politicisation Process

The Tunisian Islamic movement's first real base was at the University of Tunis. As we have seen, young enthusiastic students responded with zeal to the Islamists' message, and congregated in the colleges and faculties of what was then the sole university in Tunisia, concentrating themselves mainly in and around the capital. It was there that what Ghannouchi has described as "the struggle between Islamists and Marxists"[1] began.

Conflict on Campus

Although Marxist elements had dominated the Tunisian student movement since the late sixties, they had been unable to gain control of the official General Union of Tunisian Students (UGET) owing to governmental opposition. This led the UGET to split at its eighteenth annual congress in 1971 into pro-government and Marxist alignments. The Marxists then formed a new temporary committee to prepare for an extraordinary congress, a goal they failed to achieve throughout the ten years that they claimed to be the "sole representatives" of Tunisia's students (1971–1981).

The university campus had hitherto been the stage on which the theoretical aspects of Marxism were played out, often in the most extreme manner. Atheism was the mood of the era, French was the language of public meetings and discussions, and it was very rare to find a student who observed the daily prayers or who fasted during Ramaḍān. It was what has been described as the era of "*la gauche laïque.*"[2]

Initially, Islamist students concerned themselves mostly with opening small mosques in each faculty and dormitory, from which they organised their meetings and discussed how to Islamise the university as a whole. The books of the writer Sayyid Quṭb of the Egyptian Muslim Brotherhood, who was executed by Nasser in 1966, exercised a strong ideological influence on the young Tunisian Islamists at that time. The result, according to Ghannouchi, was a surge of

> enthusiasm for Islam and disapproval of the current situation, traditional culture, the West and all that was around generally. What helped those ideas to be seen as absolute truths were the corrupt system and the mediocrity of Islamic knowledge in our society, and the scarcity of graduates in the humanities compared to those with technological and scientific backgrounds.[3]

Driven by the desire to carry the message of Islam to their colleagues, Islamist students attempted to participate in the public debates on campus, organised usually by the Marxists. The latter were unable to find a place in their ideology for a "feudal" theory which they thought had been buried long ago. Indeed the Marxists clearly insisted, as one of their leaders wrote, that the Islamic trend was "ideologically an obscurantist movement" and "one of the most reactionary contemporary religious movements" at that time. It was argued that this was due to "its opposition to all scientific thought, and its understanding of the world built upon reactionary, metaphysical and idealistic concepts."[4]

Inevitably, peaceful cohabitation between the two movements was difficult to achieve, but there was, none the less, a positive result to this interaction, as Lamchichi explains: "Out of this ideological but very often physical and violent, confrontation between Islamists and leftist movements, Islamic movements were to learn methods of action specific to the secular left: political meetings, mural newspapers, clandestine magazines, debates, analysis and polemical politics."[5]

The leftist groups combined their efforts on 26 December 1977 to warn the currently small group of Islamists not to challenge their political dominion. Armed with knives, they attacked a general meeting that had been organised by the Islamists in the faculty of science. The police did not intervene and a number of Islamists suffered grave injuries. The last major attack which the Marxists succeeded in carrying out was on 30 March 1982, against the Islamists in the faculty of arts in Mannuba, in the suburbs of Tunis. Although dozens of Islamists were injured, they succeeded in capitalising on the incident by bringing in the media, which reported what had happened and published various condemnations of violence by different national parties and writers.

By resisting the violence of the Marxists, who still believed that revolutionary violence against "obscurantism" was the best way in which to maintain their tight grip on the student movement, the Islamists were also "intellectually" affected. Here follows Ghannouchi's account of what took place:

> The Islamist students' movement fought ferocious battles against the Marxist forces, which had a complete system with which to analyse the present and the past and to plan for the future, against the backwardness of contemporary Islamic thought in these fields. This led the Islamists to absorb a great deal of the Marxists' views in the midst of the fight against them. Those views were made easier to absorb by similarly revolutionary Iranian ideas, which presented vague concepts which could be filled by Marxist theories. All this led to confusion in the fields of

> thought and behaviour. Although these modern views enhanced the importance of political and social matters, they made Islam politically and socially militant, aiming all its arrows towards its opponents from other classes and parties, rather than preaching compassion and mercy towards people, including one's opponents.[6]

These disturbances were to prove very significant in the history of the Tunisian Islamic movement. Students were gradually expressing more interest in political, rather than merely religious, issues. At the same time, their politics were becoming more and more radical in their opposition to the régime, as they attempted to match the radicalism of the leftists. Representing a majority in the movement's membership, the students in fact forced the leadership to increase its interest in politics, and, to a certain extent, allow them to share the responsibility of directing the movement. Here is Enneifer's testimony about the situation in the movement both before and after the emergence of the student contingent:

> [Before] our ideological project was unclear; we used to speak about the first centuries of Islam, but that was all only very general.... [Things] started to become more precise when the first pupils from secondary schools passed into the university. This brought them into contact with people who were different from those of the groups of the left and the extreme left. The university thus played a central role. Since 1977 we can even say that there was a reversal of roles: the university became the locomotive of the movement.[7]

It should be mentioned here that there was one important factor which served to assist the Islamist students. Within their faculties they were free to function as a political movement, therefore they were able to oppose the régime openly and publish official statements signed in the name of *al-Ittijāh al-Islāmī*, commenting on domestic and international issues. They were even able to mock the president, and to call for a revolution to bring down his régime.[8] Because the government had in fact lost all hope of regaining direct control over student politics inside the university, the only opposition to the Islamists therefore came from the left, which helped the Islamists to win more sympathy from the student body.

Enjoying their freedom, the Islamist students became increasingly politicised, and began to demand that the entire movement move in that direction. As Hermassi explains, they were keen to address what they saw as the "general crisis" in Tunisian society, being a state-wide crisis affecting the political, economic and cultural sectors. For these young people, facing moral and spiritual laxity and bleak future prospects, their condemnation extended not only to the state but even to society, which they condemned as a *jāhilī* society.[9]

Instead of the predominantly religious and cultural issues that Ghannouchi, Mourou, Enneifer and their colleagues discussed in their lectures or articles in *al-Ma'rifa,* the Islamist students were more concerned with expressing their outright rejection of the régime and propagating anti-imperialist literature, espousing a social view supporting labour forces against capitalists, and freedom fighters around the world. For every position taken by the Marxists, the Islamists insisted on proposing an alternative. In this context the students may be seen to have played a significant role in politicising *al-Jamā'a al-Islāmiyya,* and compelling its leadership to become much more involved in the political affairs of the country.

The Government's Role in the Emergence of the Movement

As a result of these factors—the influence of *al-Tablīgh* movement and the Muslim Brotherhood, and the struggle against the Marxists at the university—the Tunisian Islamic movement gradually took shape. For some historians and analysts, however, the emergence of *al-Jamā'a al-Islāmiyya* was not such a straightforward development, arising simply in response to economic failures and extreme secularism. Rather, they suggest that the Islamic movement was supported and encouraged, if not invented, by Bourguiba's régime itself, as a means by which to oppose the rising leftist groups in the seventies. Here is Mohsen Toumi's version of events:

> It was in its capacity as a religious authority that the Tunisian government, from 1970 [onwards], began to encourage Islamist demonstrations. It assigned to Islamists a prime target, the University of Tunis, to counter the leftist students there, whether they were leftists from the ruling Destour party or communists from the extreme left. It required them also to confront the left in all social fields, and gave them the means [by which to do so]. From the beginning of the decade, the first associations for the preservation of the Qur'an appeared. In 1971 the first Islamic circles appeared in Tunis's mosques. In 1972, *al-Ma'rifa* magazine went on the market in a new, well-printed format. Clashes in the faculties multiplied. Gradually the Islamists' demonstrations spread from the university ... to the entire Tunisian society.[10]

This version reduces the entire history of the Tunisian Islamist movement to little more than a simple plot by the régime, with which to counter the leftist groups. What is more, this theory was a cornerstone of Marxist propaganda both inside and outside the university, in an attempt to discredit the Islamists and refute their claims of being opponents of the government. The leader of the communist labour party, Ḥamma al-Hammami, spoke of an alliance:

> The Islamists' hostility towards leftist groups ... was the basis of an alliance with the government to fight the influence of those groups, especially among the youth, and to absorb social discontent and redirect it in a different direction, and in doing so avoid making radical changes to the status quo. On this platform, the govern-

> ment arranged numerous facilities for the Islamists: mosques were open to their activists from which to give lectures from their highest podiums, while the press belonging to the ruling party helped them to publish *al-Ma'rifa* and *al-Mujtama'* magazines, which specialised in fighting progressive thought and publicising obscurantist views.[11]

There is a sharp contrast between this view and that of Hermassi, who, as we saw in chapter 1, stressed that "the Islamist movement thrived on the ideological looseness of the post-independence state," adding, "[it] has had the capacity to attract many more of those left behind by economic growth, or to talk like Max Weber, of the proletarian intelligentsia who despair of the nationalist and socialist discourse."[12]

Here Hermassi analyses the Islamist phenomenon from the perspective of a sociologist, whereas Toumi and al-Hammami have both attempted to address the issue from a political, rather than a historical, point of view. What is more, they have attempted to offer a feasible explanation to the question that has haunted many young Marxists in Tunisia and elsewhere in the Arab world: how could Islamism become the strongest ideological and political force of opposition to the régime in place of the Marxist forces? After all, according to Marxist rhetoric, Islamists are "allies of the exploiting classes and enemies of history's progress."[13]

The Islamists, of course, refute these accusations concerning their "co-operation" with the government.[14] Facts mentioned earlier, such as their dismissal from the Association for the Preservation of the Holy Qur'an in as early as 1971, and their problems in Sousse with the police in 1973, indicate that there was no such co-operation. Rather, it may be safe to assume that the authorities did not initially realise the potential threat of these young Islamic activists. The reasons for this are twofold: in the late sixties the policies of Bourguiban modernism were seen by many as highly oppositional to traditional and contemporary Islam, thus precluding any chance of an imminent surge of Islamism. And then there is another factor—a rule common to the politics of the Arab world in general—being that the first priority for every régime is to stay in power. In order to achieve this, it will resort to all sorts of repression against any opposition movement that threatens to take power and change the political system. However, if an opposition movement is seen to be weak and incapable of threatening the status quo, it will not necessarily be repressed.

In the seventies, when the Islamists formed their first circles, all indicators suggested that Islamism as a social and political factor could not re-emerge to play an important role on the political scene. Therefore the government did not feel threatened by Ghannouchi and his colleagues, thus giving the Islamic movement the chance to grow and improve its methods of work in relative peace and security. Indeed the only serious enemy recognised by the government in the seventies was the leftist movement, whether in the university, or in the trade unions, or even within certain circles of the ruling party itself, especially those who were still sympathisers of Ben Salah's socialist vision. Thus the security forces concerned themselves largely with addressing the Marxist threat, and to that end they arrested hundreds of leftist militants and sent them to prison, or into the desert for military training.

When the political climate changed dramatically in the late seventies and at the beginning of the eighties, the government was obliged to change its rules on security: the Islamists became "the enemy" and measures against the Marxists were accordingly relaxed. When Ghannouchi and the entire Islamist leadership were sentenced to long terms of imprisonment in July 1981, Bourguiba gave an audience to Mohamed Harmal, the leader of the Tunisian communist party, and gave him authorisation to operate as a legal party.

Beside the fact that the government first underestimated the potential political threat of the Islamists, and that its security forces were busy pursuing the leftists, there is a final point worth mentioning here: the emergence of Islamic political movements and the discourse of political Islam in Tunisia were not isolated cases. Indeed, during the seventies it was a prevalent phenomenon in the entire Islamic world, notably after the release of the Muslim Brotherhood leaders in Egypt.

The Impact of the Liberal Experiment

Another important factor that pushed the Islamic movement towards politicisation came in 1978, again in response to the negative effects of economic and social failures. The policies of Prime Minister Hedi Nouira, which had initially seemed as though they would be capable of rectifying the problems created by Ben Salah's programme, still failed to satisfy the needs of large sections of Tunisian society. This time, their anger could not be repressed.

Nouira (1911–1993) had been an active member of the Destour party both during the struggle for independence and afterwards. Like Bourguiba, he was born in the coastal city of Monastir, and also graduated as a lawyer from Paris. He returned to Tunisia in 1937 to take on an active militant role in Bourguiba's party and in the trade unions. In the first government formed after independence he was appointed finance minister; he also became the first governor of the Central Tunisian Bank in 1958. During the sixties he had been somewhat unenthusiastic about Ben Salah's socialist departure, although he succeeded in keeping a low-profile post in the government because of his close relationship with Bourguiba. Thus, when Ben Salah failed, he was soon to emerge as the saviour, and was able to implement the solution he so strongly supported: a market economy.[15] Nouira argued that it was mainly for the private sector, Tunisian or foreign, to stimulate the economy, and not for the state alone.

Various incentives for investors (especially exemptions from the payment of taxes) were approved in two decrees passed in April 1972 and August 1974.[16]

However, this period of economic revival turned out to be short-lived: the private sector invested mainly in the textile and construction industries, and was unable to fulfil the rising demand for jobs. Wealth continued to be concentrated in the hands of a minority, and poverty again increased.

In 1975, an official report showed that while 7.6 percent of the population spent more than TD400 per person per year, 32.5 percent spent less than TD80. Also, foreign debts rose by 50 percent between 1972 and 1976, from TD392 million to TD588 million.[17] Salaries were effectively frozen for the large number of civil servants in education and public administration.

Employees affected by the new hardships soon resorted to strikes, as a means by which to vent their anger. The government found itself unable to tolerate these manifestations of popular protest and the situation deteriorated markedly, to the point that the executive committee of the Tunisian General Workers' Union decided to call a general strike throughout the entire country on 26 January 1978. This day was to go down in Tunisian history as the "26th of January Uprising," or "Black Thursday."

The economic plan of Hedi Nouira had alienated large sections of the middle class who, according to Hermassi, had traditionally been long-time supporters of the régime. Thus the government lost a large part of its support base, and a mood of low confidence prevailed in the economic sector, which had been overrun by opportunists. This mood soon spread to the education sector; education as a value and teachers lost respect, and "relations became based on gain and interest."[18]

Young graduates also began to articulate their dissatisfaction, again weakening government support, and on the morning of Black Thursday pupils from secondary schools and university students joined the workers in their general strike. They took to the streets *en masse*, while the army was called in in an attempt to restore order. The result of the confrontation was graver than anyone had anticipated: according to official statements 51 protesters died and some 400 were injured, whereas Toumi has estimated figures of around 200 dead and at least 1,000 injured.[19]

The consequences of the general strike were such that it was almost impossible for anyone in Tunisia to be indifferent towards or unaffected by the turn of events. Speaking about the reactions of the Islamists, Ghannouchi writes:

> We remained on the sidelines of these violent events and did not take part in any of them. We had no unionist activity because we were somehow prejudiced [in thinking] that unionism was alien to us. Social confrontation between rich and poor is a Marxist formula which did not correspond to our understanding of life. Later, we realised that Islam also has a say in that confrontation and that as Muslims we could not stay indifferent to it. Islam gives support to the oppressed. From that point we began to develop a consciousness and sensibility towards social realities.

> The Islamists started to participate in the trade union movement's activities and nowadays represent a very powerful force in this field.[20]

As is admitted here by the movement's leader, and as will become clear in the charting of its recent history, *al-Jamā'a al-Islāmiyya* was to a large extent the fruit of its founders' experiments, trials and errors. Since none of its founders had formulated a clear social or political Islamic theory, and none could be truly described as a religious scholar, their course of action was to rely largely on their energies and their reactions to events around them—in the university, in society, or in the greater Islamic world as a whole. Thus the events of the January Uprising served as a catalyst, and compelled the Islamist movement to publish its first political statement. According to Enneifer, "it was an unsigned statement, drafted by Ghannouchi and reviewed by us together. It was entitled 'Before the Fall of the Iron Curtain', meaning that at that moment we were primarily concerned that the country might fall into leftist hands. We thought Islam in Tunisia was in danger and that the left was going to seize power."[21]

The statement not only supported what it called "the legitimate demands" of the workers, but also condemned the violence of the demonstrators and warned against being "dragged into a civil war in Tunisia."[22] This statement marks the starting-point for more political participation on the part of the Islamic movement. Because it was under the influence of those Islamist students trying to match the revolutionary stand of their Marxist opponents, the movement's position towards the régime became increasingly radicalised. It shifted from taking a purely religious stand, which condemned the government only for its "anti-Islamic" policies, to what Ghannouchi has described as "a comprehensive theological, political and social view which condemned the régime's dictatorship, alliance with foreign powers, Westernisation and exploitation."[23]

Very soon after the events of 26 January 1978, the movement began to infiltrate the trade unions and establish a solid base there. It also began to raise the issues of social justice, the rights of workers, and an Islamic theory of development. In 1980 the movement recognised for the first time the activities of 1 May International Labour Day, and arranged for a special celebration, organising a large meeting in one of Tunis's mosques to discuss social theory in Islam. After 1978, Islamist students were not alone in their competing with the Marxist groups; Islamist trade unionists were obliged to do the same within the trade unions, and the movement's leadership was forced to adjust its message to meet the new needs of the moment.

The Islamists realised by this point that their movement had to, as Hermassi notes, "wisely reconsider its direction. Time was no longer for unrealistic thoughts about the absolute state and pure Islamic justice, but for building direct relationships with the real people and their problems and interests as defined by the people themselves," and argues that their inclusion with the trade unions in the late seventies woke up the Islamists from their "dogmatic sleep."[24]

These were the new challenges facing Ghannouchi and his colleagues after the crisis of 1978. Accordingly, they were obliged to widen their frame of reference in order to seek the answers with which to meet these new challenges. They were further inspired when just one year later they witnessed what had formerly seemed unthinkable: the Islamic revolution in Iran and the ascendance of an Islamic state.

The Iranian Factor

When Ayatollah Khomeini's plane finally landed in Tehran on 1 February 1979, ending 15 years of exile in Turkey, Iraq and France, the joy and excitement among the Islamists were unanimous. Many of them had already given up hope that Iran would continue to be an Islamic country, even when compared with the model of secularised Turkey. However, the political opposition that had built up in Iran during 1977 and 1978 came to a sudden and unexpected conclusion in the February of the following year: the Shah had fled the country, to be joined very soon by his Prime Minister Shapour Bakhtiar, and the popular revolution had won the struggle under the leadership of a scholar who had worked patiently for his aim to build an Iranian Islamic state.

For the majority of Islamists throughout the Arab world, the Islamic victory in Iran was "a dream come true"; a distant goal to which they had long aspired had suddenly become a reality. It led them to believe—in the words of one of their opponents—that they were "close to victory," because they were "the soldiers of God, and God says, 'And that our forces they surely must conquer'."[25]

In Tunisia, the Islamists' support for the Iranian Revolution was genuine and overwhelming. In *al-Ma'rifa* magazine, issued that year on 12 February, Ghannouchi urged Islamic movements in the Arab world to offer their unequivocal support for the revolution: "We want a strong, clear, frank and solid stand from the militants of Islam everywhere and the supporters of freedom and justice with their brothers in religion, the fighters for freedom and justice in Iran ... those who have raised the flag of Islam high over the entire world."[26]

In August 1978, the movement had begun to publish a weekly newspaper under the title *Al-Mujtama'*, and through this organ the Tunisian Islamists expressed their enthusiasm and support for the Iranian revolution. On campus, the Islamist students finally found themselves defending an Islamic, anti-American revolution, after having been labelled for so long by the Marxists as reactionary, pro-American elements.

Khomeini was seen by the Islamists as the leader of this onslaught against the United States, defending the oppressed (*al-mustaḍ'afūn*), standing against the capitalists (*al-mustakbirūn*), and inciting the masses to overthrow their corrupt ruling dictatorships. Thus he served as a role model for the Islamist students in their fight against the Marxists, while his revolution supplied them with slogans and terminology. As a result, the Islamic movement on campus organised huge celebrations in honour of the events, and Iranian slogans were immediately reproduced in its statements and newspapers.

Thus we may see that the Iranian revolution also played a very important role in politicising the Tunisian Islamic movement. Hermassi believes that it "activated the overwhelming hopes of Islamists, and their ideas about taking power. The tendency towards totalitarian rule which had always existed in the contemporary Islamic trend had finally materialised. The utopia became not only acceptable and realistic, but possible too."[27] Inevitably, it was not long before optimistic young Tunisians began to ask themselves: if Khomeini had succeeded in Iran, then why could the same success not be repeated in Tunisia? As Boulby observes, "Ghannouchi glorified the revolution to his followers: 'The example of Iran', he told them, 'shows us the awakening has come. Fight against licentiousness and make sacrifices! To correct others and make our own revolution we have to correct ourselves and worship God'."[28]

In all, the most important impact of the Iranian Revolution on the Tunisian Islamists is that it supplied them with greatly needed, realistic, political positivism. Pessimism, a readiness for sacrifice, gaols and torture had been the vocabulary of their former mentors, the Egyptian Muslim Brotherhood, as a result of their confrontations with Nasser. The Islamists in Iran, however, provided a model of success and political power, which became a source of great motivation for the members and supporters of *al-Jamā'a al-Islāmiyya*.

The numbers of its followers increased dramatically. The Islamists began to represent the largest political group in many secondary schools and faculties. Those who graduated and went on to teach in primary or secondary schools were also very active in spreading the movement's message, as were other organisations throughout the country. By the end of the seventies, their support base was wider than anyone had expected, which once again caused the leaders to reconsider the movement's means and methods. As a result, they agreed to convene their first official conference.

Secrecy: The Benefits and Costs

The conference took place in Mannouba, in the northern suburbs of Tunis. Around seventy of the most prominent members of the movement convened in August 1979 at what will be called the "founding" conference. They agreed on a constitution for their secret association, giving it a more detailed administrative structure, an institutional breakdown of which will be listed here, in order of importance:

1. **The general conference**, convened once every 3 years. This represents the highest authority in the movement and decides its main policies.

2. ***Majlis al-shūrā***, (similar to a small parliament), made up of 14 elected members from the general conference. This constitutes the legislative branch of the movement, with sufficient power to overrule the executive bureau. Members meet at least once every 3 months.
3. **The executive bureau**, run by the *amīr* (president) of the movement, who is elected by the General Conference. This bureau exercises its functions with the help of various sub-committees.
4. **The *'Ummāl***, meaning the presidents of the movement in each province, parallel to official provincial governors. They are appointed by the *amīr* to run the movement in the provinces with the help of a local executive bureau and *majlis al-shūrā*.
5. ***Wukalā'***, meaning those who preside over the movement's activities in the main districts of the provinces.
6. **The university**, considered an independent entity in the movement, almost like a province. Its leadership runs the activities of Islamist students all over the country. It is also run by an *'āmil*, appointed by the *amīr*.
7. **Secondary schools**, also viewed as a separate entity, run by a central committee which plans the movement's activities within these schools.
8. **The cells**, which are the basic units of the structure. Every member of the movement should have around 3 years of training in a two-level system designed especially for these cells, before he or she can be granted the rights of a full member of the movement. The training programme consists of courses in Islamic sciences and contemporary Islamic movements.[29]

The Mannouba conference also decided that the movement was part of the international trend of the Muslim Brotherhood, and elected Ghannouchi as its *amīr*. Ghannouchi had in fact been the *de facto* leader since the foundation of the movement's first cells. In 1977 he became the editor-in-chief of *al-Ma'rifa* magazine and established a reputation for himself as one of the most prominent lecturers in the mosques of the capital. He continued to work as a teacher of philosophy at a secondary school in Tunis, and travelled abroad on a number of occasions, notably to Saudi Arabia and Sudan, where he made direct contact with Dr Ḥasan al-Turābī, the leader of the Muslim Brotherhood in Sudan.[30]

Other main figures in the elected leadership of *al-Jamā'a al-Islāmiyya* included Salah Karkar, selected by Ghannouchi as his deputy. A teacher of economy and a keen organiser, he was to become a symbol of the radical hardliners among the Islamists and later emerged as Ghannouchi's main rival in the movement's leadership. Abd al-Fattah Mourou was elected to the legislative council, but Ehmida Enneifer did not even feature in the list of the conference's members. From 1976 onwards, he had become increasingly critical of the influence of the Muslim Brotherhood on the movement, and called for a different, more rational and progressive approach to contemporary issues. It seemed also that his personal relationship with Ghannouchi had become strained, and that he was dissatisfied with Ghannouchi's leadership of the movement.[31]

Around this time Enneifer had begun to meet with a number of Egyptian Islamic writers who advocated the necessity of renovation in the Islamic movement, and had become interested in the opinions of the Egyptian Islamist Dr Muḥammad Fatḥī 'Uthmān, and his notion of an "Islamist left." Enneifer wrote a series of articles in *al-Ma'rifa* magazine entitled, "Where Do We Begin?," the focus of which was on "Islamic thought as the main source of the problems in which Muslims find themselves and as the key for effectively solving these problems."[32]

Enneifer decided shortly afterwards to resign from *al-Jamā'a al-Islāmiyya* and to form a new, reformist group under the name of *al-Islāmiyyūn al-Taqaddumiyyūn* (the Progressive Islamists), with the main task of "restructuring Islamic thought by rethinking the fundamentals of Islam."[33] This was different from the main mission of *al-Jamā'a al-Islāmiyya*, which was to restore society to Islamic rules; in other words, the problem was seen to lie with society and not with Islamic thought. Ghannouchi announced finally that his movement was unable to contain its differences with what he called "the rationalist trend." Enneifer and his supporters finally seceded and published their own magazine with the title *15/21*,[34] referring to the combination of the Islamic and Western calendars, and symbolising the need to find a new, progressive deal between Islam and the West.

It was agreed that the new structure for the movement should be kept secret, although this presented problems in that it was large and difficult to hide. Further, with growing radicalism in the movement's position towards the régime, and increased media interest in Islamic revivalism after the Iranian Revolution, secrecy was not easy to sustain. The government had set up a special branch in the interior ministry to monitor religious activities, and on 7 December 1979 Ghannouchi was arrested and questioned for 17 days, as was Mourou, who was detained for 3 days. They were asked about lectures they had given in mosques and statements they had issued against the régime, but not about their secret organisation, as yet unknown to the security forces.[35]

Al-Ma'rifa magazine was suspended in the same year, and the official media began to attack the Islamists. Even Prime Minister Hedi Nouira expressed annoyance: on 19 December he denounced "those troublemakers who used religion for their political aims."[36] With the leaders of the UGTT still imprisoned after their role in the January Uprising, the Islamic movement soon emerged as the "next main threat" to the Tunisian régime. Under close observation, the movement's success in preserving its secrecy came to a bitter and very abrupt end.

The Discovery of the Secret Organisation

The fifth of December 1980 was unlike any other day in the Tunisian Islamic movement's history. According to Ghannouchi: "It was no less important than the success of the Iranian Revolution in politicising our Islamic thought and changing the views of the vast majority of Islamists in Tunisia. We may even consider 5 December 1980 to be the separating point between two stages in the Islamists' history: the *da'wa* stage and the comprehensive political stage."[37]

Tunisian police had arrested two of the executive bureau's members: Salah Karkar and Ben 'Issa Dimni. They found with the latter detailed written documents about the movement's strategies and activities. Subjected to severe torture, Demni gave the police crucial information about the movement's organisation. After being tortured and challenged with Dimni's information, Karkar in turn gave the police a detailed description of *al-Jamā'a al-Islāmiyya*, its foundation, its institutions, the founding conference, its *majlis al-shūrā*, the executive bureau, the local bureaux, and the committees. He also gave the names of the main members of these institutions. The two men were finally released after a week, during which time the régime had unearthed what Ghannouchi describes as "a treasure trove of information."[38]

Suddenly, "the enemies," as Ghannouchi calls the régime, were fully aware of all the movement's activities and actions, leaving its direction in utter disarray. Steps were swiftly considered in an attempt to save the movement from total disaster. First, a number of technical measures were taken: the *majlis al-shūrā* decided to dissolve itself, as did the central executive bureau and committees. The question was then posed of how to continue running the movement's various activities. It was then that Ghannouchi proposed to his colleagues a dramatic solution: to expose the movement, and make the facts known to the public before they were misused by the régime. It was a solution forced by events rather than premeditation. Recalling this period, Ghannouchi admits:

> Dialogue with some brothers led to the proposition of announcing [ourselves as] a political movement at a press conference, to cover up for our exposed secret activities, especially because the régime had begun—under pressure from within and from outside—to use democratic slogans and to prepare for a phase of pluralism, even if limited to Bourguiba's terms. So [it was decided:] why don't we seize this chance in order to deprive the government of the chance to crack down on us and announce the discovery of a "secret dangerous movement"?[39]

The proposition was met with a positive response from 71 percent of the movement, according to a hurriedly organised poll requested by the leadership. The main opposition came from students and radical elements in the leadership, notably deputy president Salah Karkar. They argued that requesting legal status from the enemy would be equal to admitting its legitimacy, and that in their view the régime was corrupt, dictatorial, an agent of the West and an enemy of Islam, and that to request its permission to form a political party was "a crime no less than that of *kufr* or high treason."[40]

Finally, an extraordinary general conference was called, to decide on new tactics after the authorities' discovery of the nature of the movement. The conference was convened on 9 and 10 April 1981; at almost exactly the same time, Bourguiba's ruling party held an extraordinary conference at which it set out to tackle its own associated problems. One of the items on their agenda concerned an attack on the city of Gafsa in the south of the country, on the night of 26 January 1980, by a group of Arab nationalist militants backed by Algerian and Libyan intelligence. The date of the attack had been scheduled to coincide with the anniversary of the January Uprising; the attackers had planned to "liberate" Gafsa, as the first step towards asking for official Libyan assistance.[41]

The plan failed after the army was called into Gafsa: dozens of people were killed, and 300 were arrested, including all the members of the group. After a hasty trial 15 of them were charged with high treason and sentenced to death; they were hanged on 17 April 1980. The political message received from the attempted coup was that a change in the political system was very much needed. The government's problems were exacerbated further when Hedi Nouira was stricken with medical paralysis and was no longer able to continue in office.

This opened the way for Mohamed Mzali, one of Bourguiba's long-serving ministers, to become government co-ordinator; a few months later he was appointed as the new Prime Minister. He too supported the idea that liberalism was the best way out of the socialist failure of the end of the sixties, and now Bourguiba turned to a new slogan: openness and a new era based on a multi-party political system. At the April 1981 conference, he told his ruling party members that he no longer objected to the forming of other social, political and national associations.[42] Two weeks later, an extraordinary conference of the UGTT marked its reconciliation with the government after the release of its leaders. They were allowed to return to their posts in the organisation except for their leader Habib Achour, who would return at a later date. For opposition groups, the new era looked very promising.

The Creation of the Political Party

All these various factors from both within and outside formed the background for the Islamists' meeting in Sousse at their extraordinary general conference, as they attempted to address their crisis effectively. They first listened to Bourguiba's speech about a multi-party system, and then to Ghannouchi about the need to announce themselves as an Islamic political movement. The idea was finally accepted in a new strategy, designed to protect the movement from its main enemy and only real threat: the régime. Ghannouchi was once again elected to the post of *amīr*.

The new *majlis al-shūrā* met frequently over the following weeks to work out the exact details for this new departure and to prepare for their public announcement. A statement for the occasion was agreed upon, as was the formation of a new body called "the founding committee," which was given the responsibility of applying for the legal registration of the new party.

Its members met on 29 May 1981, at the home of one of the old famous Zeitouna scholars, Shaykh Mohamed Salah Enneifer. There they elected the five members of their political bureau: Abdel Fattah Mourou, Ben 'Issa Demni, Zahir Mahjoub and Habib Mokni, with Ghannouchi as its president, and Mourou as its secretary-general.

By June 6 1981, *al-Jamā'a al-Islāmiyya* had finally changed its name and course. Its leaders organised the holding of a press conference, which was attended by Tunisian and foreign journalists. From Shaykh Mourou's office in Commission Street, which demarcated the old city of Tunis from the modern one, Ghannouchi announced the creation of *Ḥarakat al-Ittijāh al-Islāmī* (the Islamic Trend Movement) in Tunisia. He declared that the movement wished to be recognised by the government as a legal political party, and pledged that it would play a constructive role in bringing about success for the new era of "openness," or democracy. This was the first page of yet another chapter in the history of the contemporary Tunisian Islamic movement.

Notes

1. Al-Ghannūshī, interview, *Arabia*, April 1985.
2. Lamchichi, *Islam et contestation au Maghreb*, op. cit., p. 192.
3. Al-Ghannūshī, *Ḥarakat al-ittijāh al-islāmi fī Tūnis*, op. cit., pp. 109–110.
4. Al-Hammāmī, *Ḍiḍḍa al-ẓalāmiyya*, op. cit., p. 8.
5. Lamchichi, *Islam et contestation au Maghreb*, op. cit., p. 192.
6. Al-Ghannūshī, *Ḥarakat al-ittijāh al-islāmi fī Tūnis*, op. cit., p. 110.
7. Enneifer in Burgat, *L'islamisme au Maghreb*, op. cit., p. 209.
8. Al-Ghannūshī, *Ḥarakat al-ittijāh al-islāmi fī Tūnis*, op. cit., p. 111.
9. Al-Harmāsī, in *al-Ḥarakāt al-islāmiyya al-mu'āṣira fī al-waṭan al-'arabī*, op. cit., pp. 264–5.
10. Toumi, *La Tunisie de Bourguiba à Ben Ali*, op. cit., pp. 116–7.
11. Al-Hammāmī, *Ḍiḍḍa al-ẓalāmiyya*, op. cit., p. 7.
12. Elbaki Hermassi, "The Islamicist Movement and November 7," op. cit., p. 193.
13. Al-Hammāmī, *Ḍiḍḍa al-ẓalāmiyya*, op. cit., p. 18
14. Burgat, *L'islamisme au Maghreb*, op. cit., p. 214.
15. Toumi, *La Tunisie de Bourguiba à Ben Ali*, op. cit., p. 130.
16. Ibid., p. 134.
17. Ibid.
18. Al-Harmāsī, in *al-Ḥarakāt al-islāmiyya al-mu'āṣira fī al-wa an al-'arabī*, op. cit., p. 264
19. Toumi, *La Tunisie de Bourguiba à Ben Ali*, op. cit., p. 155.
20. Al-Ghannūshī, interview, Arabia, April 1985.
21. Enneifer in Burgat, *L'islamisme au Maghreb*, op. cit., pp. 208–9.

22. Walīd al-Manṣūrī, *al-Ittijāh al-islāmī wa Burqayba: muḥākamat man li-man?* (Tunis: n.p., 1988), p. 19.
23. Al-Ghannūshī, *Ḥarakat al-ittijāh al-islāmī fī Tūnis,* op. cit., p. 39.
24. Al-Harmāsī, *al-Ḥarakāt al-islāmiyya al-mu'āṣira fī al-waṭan al-'arabī,* op. cit., pp. 278–9.
25. Muḥammad A. Khalafallāh, "al-Ṣaḥwa al-islāmiyya fī Miṣr," in *al-Ḥarakāt al-islāmiyya al-mu'āṣira fī al-waṭan al-'arabī,* p. 73.
26. Al-Ghannūshī, *Maqālāt,* op. cit., p. 83.
27. Al-Harmāsī, *al-Ḥarakāt al-islāmiyya al-mu'āṣira fī al-waṭan al-'arabī,* op. cit., p. 266
28. Boulby, "The Islamic Challenge," op. cit., p. 603.
29. Al-Manṣūrī, *al-Ittijāh al-islāmī wa Burqayba,* op. cit., p.16.
30. Burgat, *L'islamisme au Maghreb,* op. cit., p. 213
31. Al-Ghannūshī, personal interview, 23 May 1993.
32. Magnuson, in Zartman, ed., *Tunisia: The Political Economy of Reform,* op. cit., p. 180.
33. Al-Ghannūshī, *Ḥarakat al-ittijāh al-islāmī fī Tūnis,* op. cit., p. 35.
34. Ibid., p. 139.
35. Burgat, *L'islamisme au Maghreb,* op. cit., p. 209.
36. Al-Ghannūshī, *Ḥarakat al-ittijāh al-islāmī fī Tūnis,* op. cit., p. 136.
37. Ibid., p. 139.
38. Ibid., pp. 143–4.
39. Ibid.
40. Sophie Bessis and Souhayr Belhassen, *Bourguiba: Un si long règne (1957–1989),* Vol. 2 (Paris: JAPRESS), 1989, p. 168.
41. Toumi, *La Tunisie de Bourguiba à Ben Ali,* op. cit., pp. 159–60.
42. Bessis and Belhassen, *Bourguiba: Un si long règne (1957–1989),* op. cit., p. 18.

3

Islamists v. Bourguiba: 1981–1987

The day when Tunisia's *al-Jamā'a al-Islāmiyya* announced to the local and international media that it wanted to be recognised as a legal political party functioning within the bounds of the constitution, was a turning-point in the history of the Tunisian Islamic movement. Ghannouchi says of the occasion:

> It was as though I had disposed of a mountain that my back could not carry. We were in a race against time with the régime, and knew that the documents accusing us were ready and that they would be used against us the moment the higher authorities decided.... [However] I was so happy that even if I were to have been hanged after that I should not have minded.[1]

The days of secrecy had come to an end, and from 1981 onwards *al-Jamā'a al-Islāmiyya* was compelled to engage in a very delicate and dangerous political game: on the one hand it had to maintain its complex underground organisation, while on the other it had to run a political party with its different obligations. Each of these two organisations had its own particular needs, personnel and rhetoric. It proved to be a difficult task to undertake.

As Ghannouchi has admitted, the formation of *Ḥarakat al-Ittijāh al-Islāmī* as a political party was forced upon the movement and, initially, the new party was unable to propose any clear manifesto. When challenged on this point, the leaders of the movement argued that harassment on the part of the authorities had not allowed them to draft such a manifesto. Ghannouchi has described this scenario accordingly:

> Both the leadership and the members had tried to draft and agree on an ideological approach and a strategy for action. We spent a long time [in doing so] and prepared large files of theoretical studies on Islamic social views and the current situation in the country from a variety of aspects. Special committees had begun to draw their final conclusions from these studies ... but as soon as we had begun, the events of 5 December 1980 took place and the police seized all our files, destroying the efforts

of more than a year, and forcing the movement to concentrate on defending itself rather than improving its theories.[2]

For those demanding a detailed manifesto this excuse was unacceptable. Indeed, one Tunisian critic, who published an entire book in an attempt to "confront the message of ignorance,"[3] argued that the absence of the Islamists' programme was deliberate, because "it enabled their leaders to manoeuvre freely and escape supervision by and accountability to the movement's members. It also enabled them to attract a greater number of supporters, if we consider that religious thought is still the ideological framework for society as a whole."[4]

The new party did, however, present a long statement declaring the founding of the movement and outlining its ideological and political platform. It declared that Islam had been "the target of all conspiracies waged by tyrannies inside Islamic countries, or by Western imperial powers from outside," and added that it had been "marginalised in directing and ruling the day-to-day life of Muslims, despite the fact that it was Islam which was behind the success of the Islamic civilisation and behind the success of the national struggle to gain independence from European colonialism."[5]

Under the banner of re-establishing Islam as a dominant factor in ruling Tunisian society, the statement specified five major objectives for the movement:

1. To revive the Islamic personality of Tunisia so that she may resume her traditional role as a major centre of Islamic civilisation and culture; this would necessarily entail putting an end to the wastage of national resources, the progressive alienation of the population by the government and the practice of slavish imitation of the West.
2. To reformulate Islamic thought, taking into account the fundamental principles of Islam, the requirements of man's continuous evolution and changing circumstances.
3. To reassert popular will as a political force, and in so doing to reject internal paternalism and foreign influence.
4. To establish a system of social justice based on the principle that, although everyone should have the right to benefit from his own endeavours, subject to the public interest, everyone should also enjoy the right to receive what he needs: to each according to his efforts and to each according to his needs, so that the masses acquire their legitimate right to live in dignity, removed from all forms of exploitation and submission to international economic powers.
5. To contribute to the revival of the political and civilizational unity of Islam nationally, regionally, within the Arab world and internationally, thus saving our people and the whole of humanity from psychological alienation and social injustices and international imperialism.[6]

A separate part of this document identified the means by which these rather broad goals could be achieved, although they themselves were also very general. Suggested methods included using mosques as centres for the general mobilisation of the masses, encouraging various cultural activities that would help to enhance Islamic values, supporting efforts to make Arabic the official language of the administration, accepting political pluralism and democracy, and rejecting political violence.[7]

It may be argued that this vision was not inappropriate for an ordinary, cultural, non-political association, but to Tunisian politicians at the time, such "moderate" talk was irrelevant. What really mattered to President Bourguiba's régime was that a religious movement was now challenging him for power, a fact which he was not prepared to accept. Therefore, only five weeks after presenting their new party to the media, the leaders of *Ḥarakat al-Ittijāh al-Islāmī* (also known as the *Mouvement de la Tendance Islamique*, or MTI) found themselves in prison. On 18 July 1981 the authorities launched a wide-ranging campaign which ended in the arrest of 107 of the Islamic movement's activists, including Ghannouchi and Mourou. They were officially charged with the following "offences": the operation of an unauthorised association, the distribution of tracts, and the dissemination of false news. On 4 September they were sentenced to various terms of imprisonment ranging from six months to eleven years. After appealing, some sentences, such as those of Ghannouchi and Mourou, were reduced slightly from eleven to ten years.

The First Confrontation: 1981–1984

According to the public prosecutor's report, the main danger that had emerged in the MTI's short history was that "the movement, which had started as a purely *religious* movement, had become political, establishing its institutions and claiming legal recognition … on the basis that Islam mixes religion with politics."[8]

Using religious slogans for political ends was never accepted by other sections of the Tunisian élite within the ruling party, nor was it accepted by the main opposition groups such as the MDS (Mouvement des Democrates), the Communist Party, or the MUP (Popular Union Movement), all of which had already been granted legal permission to operate as political parties. The government was adamant that the MTI was a blatant threat to society. Its official media denounced "the entire Islamic movement, making reference to the adherents of violence, fanaticism and intolerance, and to those marginals who use religion to soil the distinguished image of the country."[9]

The accusation of violence was mainly linked to the wide unrest that had erupted among students in the secondary schools and faculties of the University of Tunisia during the spring of 1981. The Islamists had been very active in leading strikes and demonstrations around the country, to the point that they had once taken hostage the Dean of the Faculty of Sciences in Tunis in order to stop the police attacking them.[10]

The student wing of the MTI had been greatly influenced by the events of the Iranian revolution, including some of its more extreme methods of protest, such as the seizure of the American embassy in Tehran in 1979. At certain stages during that period of unrest, the movement's leadership was unable to restrain the students from demonstrating in the streets and fighting the police. It was not until the incident of taking the Dean of the Faculty of Sciences hostage that the Islamist students and their leaders in the MTI realised the extent of the harm they had done to their cause. The incident was widely condemned by all the teachers in the university and by all the political parties. The police began to pursue the leaders of the movement's student wing, and their photos were published in the press as "wanted criminals." Ghannouchi admits that "members of the Islamist Trend in the university and the whole movement in general went through a terrible crisis because of that stupid, irresponsible act."[11]

Despite these incidents, which served only to validate some of the accusations against the MTI as a potential source of violence, the real discord between the newly proclaimed party and the government and other opposition parties concerned political power, and whether the Islamists should be allowed to gain it by benefiting from their public commitment to Islam. This was manifested clearly in an argument raised by the Islamists' opponents: that the MTI should not be allowed to claim Islamic identity in an Islamic society. Toumi explains it accordingly:

> The Islamic movement does not recognise any religious legitimacy for the government nor for the other Tunisian political parties, despite certain tactical alliances. Because it claims it is the true guardian of the strict teachings of Islam, the MTI presents itself, in fact, as the sole representative of the Truth. Such thinking must definitely lead to the kind of *stato totalitario* envisaged by Mussolini.[12]

The problem in 1981, however, was that the government allowed neither the time nor the opportunity for examination of the arguments and counter-arguments about the claims of the Islamists. Repression was deemed to be the strongest response to the MTI's application for official recognition, although this line served only to alienate other political parties opposed to the Islamists, and in fact forced them to show some degree of support for this new wave of political prisoners. In fact, the period of the first confrontation proved to be a very prosperous era for the Islamists, as al-Manṣūrī has noted:

> The period of repression was a bright page in the history of the movement and the country. During it, the movement reaffirmed its unity, insisted on its legitimate rights to official recognition, and confirmed its commitment to peaceful and de-

mocratic means of political action, as well as its denunciation of violence, especially since the movement had been a victim of this both inside and outside the university. This brought respect from different political, human and professional organisations, and people's trust in the movement's sincerity. All groups supported it in demanding a general amnesty and the right of the movement to operate legally. The movement widened its popular base, and gained a strong and international reputation as a moderate group, working to make changes by democratic means, within the bounds of law, resisting the government's repression and avoiding being dragged into violence and the logic of force.[13]

We may see here that the MTI confidently presented itself in the role of the oppressed, thereby securing the sympathy of others. This was made possible because of the political climate in the country at that time, which was still moderate and even tolerant to a considerable extent, compared with earlier periods of Bourguiba's régime and with other neighbouring countries.

A few months after the arrest of the MTI's leaders, the first multi-party general elections took place. Although the ruling party won all the seats, the political debate generated by the occasion was new to Tunisian political life, and independent and opposition newspapers gained respect and attracted a bigger readership. It was in this climate that different political organisations, writers and personalities were able to speak out and show support for the MTI's political prisoners. Had the movement chosen to play the role of the party of revolution and radical change, it would not have been so successful in attracting such a level of national support. Similarly, had the régime tightened the margin for free expression, the Islamists would have considered violence as the only useful way in which to defend their existence, as was to occur a decade later.

During the early eighties Ghannouchi appeared to be happy with the outcome of the current confrontation with the authorities, as he observed it from his prison cell:

> Our arrest was a chance to feed new blood into the movement, thereby preserving its youth, effectiveness and renewal ... the movement became the centre around which all popular and political trends came together, either sincerely or for political calculations—except for the ruling party, despite the fact that some of its wings were against repression and some expressed their support for us.... Almost every week after our arrest, independent and opposition newspapers published petitions condemning the government's stand against us and demanding support for us. This led to a national demand for a general amnesty adopted by the entire opposition and some members of parliament, a demand for which *Al-Ra'ī* newspaper gathered the support of twenty thousand citizens.[14]

This assessment may in fact be somewhat exaggerated, but more than anything else, it does show that Bourguiba's régime was not fully committed to embarking on a widespread campaign to eliminate the Islamists from society. The slogan of the era was one of openness rather than dictatorship. As we see from Ghannouchi's quotation, not all the leaders of the ruling party supported policies of repression, and it seems that even the current prime minister, Mohamed Mzali, was more concerned with democratising the country than fighting the Islamists. Despite the rhetoric he had employed during the summer of 1981 against the Islamists, Mzali was seen widely as a supporter of Arab culture, and was in fact suspected of being lenient towards Ghannouchi and his colleagues. A few years after leaving office, he wrote of the MTI, which had by that time changed its name to *al-Nahḍa*:

> *Nahḍa* means "renaissance." It is very clear that during the last few years, we have been witnessing an Islamic revival in many Islamic countries, which shows that our soul is alive and active. I know by experience that *al-Nahḍa* is mainly composed of men of values, peaceful democrats who do not condone violence, and of Tunisians attached to their country and its cultural and spiritual values. There is no ground for the slightest form of indictment.[15]

Not surprisingly, the MTI was able to capitalise on its circumstances and gain unprecedented publicity for its ideas and leaders. In addition to enjoying the support of many politicians and newspapers in the country, the university continued to serve as an alternative base for the movement, for those students supporting the movement were still free to operate in the form of a mini political party. Under the banner of *al-Ittijāh al-Islāmī,* they were free to organise political meetings, to defend their detained leaders, to express their views on all national affairs, and to issue political statements that could be published by national newspapers. Islamists may not have been the majority in the university, but their presence was formidable, as Waltz has noted:

> Especially by the early 1980s, the University of Tunis had emerged as the principal breeding ground of Islamism, and within that confined arena the Faculty of Science appeared particularly fertile. Some estimates put the number of adherents as high as twenty percent of that institute's student body, but even that figure underestimates their impact. While the Islamists are clearly not a majority, they are nevertheless the best organised and the most unified group among the students.[16]

However, as was the case within the public realm, the MTI still had to face the consequences of repression against its inner organisation. First, a new *amīr* was elected by the *majlis al-shūrā,* namely Fadel al-Baldi from the city of Bousalim in the north-west of the country. In his early thirties, his main credentials were the role he had played in forming the movement in its early days and a heavy prison sentence passed on him *in absentia.*[17] After a few months of leading the movement from underground inside the country, he fled to

France where he claimed political asylum. In 1982, Hammadi al-Jebali, an engineer from Sousse, was chosen as the new leader to replace Baldi, for the operating inner *shūrā* council insisted that the *amīr* should be inside the country.

All these changes were performed in secret, and remained so until 8 January 1983, when the police uncovered and arrested most of the new leadership of the student movement of the MTI at a meeting in a flat in Tunis. The detainees in turn gave leads to the new secret national executive committee, and although Jebali himself was not arrested, a number of his more prominent colleagues were rounded up. Again the movement was forced to make public some of its secret arrangements: it issued a statement on 19 January 1983 announcing the formation of a new five-man executive bureau led by Jebali.[18]

The new detainees were sentenced on 27 July 1983 to short prison terms of up to two years, although many of them were released on the same day because they had already served their six months' term. Jebali was sentenced *in absentia* to two years.[19] The official mood was now clearly different from that of 1981, and many of the discussions between the imprisoned Islamists turned to a possible reconciliation with the government, although opinions were divided about the régime's intentions towards the Islamists and the conditions—if any—that the movement should lay down before seeking an agreement.[20]

For Bourguiba and his ministers, however, there were other, more urgent, problems to attend to. At the top of the agenda were the violent demonstrations held between 29 December 1983 and 4 January 1984 in the main cities of the country, in protest against the sharp rise in the price of bread. Although the police responded by using force, they were unable to contain the situation, and the army was called upon to help restore order after a state of emergency had been declared. Fifty people were killed before Bourguiba finally decided to abolish the new price increases.[21]

The priorities of the government thus changed from fighting the Islamists to the restoration of its credibility and the country's stability. In this context Prime Minister Mzali proposed to President Bourguiba the idea of passing a presidential amnesty on Islamist prisoners. He embarked first on secret negotiations with them via intermediaries, later negotiating directly with Abd al-Fattah Mourou, the secretary-general of the MTI, who finally sent Bourguiba a carefully written letter confirming the noble motives of his movement and its commitment to the rule of law. Finally, on 4 August 1984 Bourguiba issued his presidential amnesty, and the leaders of the Islamists were free again, although—as would prove three years later—not for long.

According to Ghannouchi, Bourguiba did not deserve to be thanked for this gesture. He argues that his release was purely the result of the government's weakness: "In January 1984, the Tunisian government went through a major crisis. The bread crisis revealed the incompetence of the régime and its politics, and consequently the government was forced to grant the people some concessions. Thus we were released by the favour of Allah the Merciful, and amnestied."[22]

The truth, however, was not so simple. The discussions mentioned by Ghannouchi earlier about the possibility of reconciliation with the government were also taking place on a wider scale outside prison. Many Islamist students did not favour the idea of a dialogue with the government, and blamed Mourou in particular for undermining the movement's principles and recognising the legitimacy of Bourguiba's régime. In an interview given later to a Kuwaiti magazine, Mourou demonstrated the lengths to which his movement was forced to go to meet the government's requirements: instead of emphasising the radical differences between the movement and the régime, he chose to concentrate on the common ground between the two sides, and the peaceful, moderate and rational nature of the MTI.[23]

During these three years of repression the main corps of activists was dispersed in gaol, in hiding and in exile. Not all were convinced of the merits of these concessions which had helped them to gain freedom; in fact, each of these groups suffered very serious internal problems, the most serious of which was between Ghannouchi and Karkar in prison.

As we have seen, Karkar had previously been Ghannouchi's deputy president, and had begun to insist that Ghannouchi's leadership must come to an end—of which fact Ghannouchi was well aware. Ghannouchi has recorded his exchanges with Karkar concerning the challenges which lay ahead for the movement, and notes that Karkar asserted that the main challenge was to find "a management team and leadership capable of taking control of things, building a good platform for real and collective action, and making good use of Islamic capacities at all levels," adding that the movement's leadership before the period of imprisonment had "proved quite clearly its inability to achieve those objectives."[24]

Karkar's remarks were intended as nothing less than a complete condemnation of Ghannouchi's leadership, and his negative feelings towards his rival were known to many. What is more, it was also known that by condemning the movement's leadership, Karkar was promoting himself as a stronger alternative. It is true that once in prison, Ghannouchi was a much weaker leader, and he has not attempted to defend his personal record, although he points out that he chose to minimise the problems of leadership in order to concentrate on what he saw as a more complex issue:

> It is the problem of a nation trying to find its way through a dense forest ... looking to find its lost civilisational identity, through its own renaissance ... looking for a social, human and civilisational model that reflects its religion and heritage and answers the current needs and challenges.... This is the main challenge, and all other managerial and technical difficulties are simpler problems.[25]

Those who had fled to France also experienced in-fighting, mostly consisting of personal conflicts between the passive, moderate Fadel al-Baldi and the more radical students that had remained loyal to their views formed during the late seventies. The two groups even went so far as to publish two small rival publications, each of which claimed to represent the movement in exile. Further, on a personal level, Baldi was unhappy with the manner in which he had been replaced in the leadership by Hammadi al-Jebali.

In August 1984, a presidential amnesty made it possible for the activists of the MTI to try to reconcile their differences and plan for the future. Freed from prison, out of hiding and back from exile, they were all poised to seize the chance to better defend their views and ideas. Two crucial questions demanded a collective answer: who would be the new leader, and what direction should the movement take; should it play a greater part in politics as a political party, or should it go back to the days of *da'wa*, as a strictly religious organisation?

The Truce of 1984–1986

Before the end of 1984, the MTI succeeded in organising its third national conference. After having gone through a wide range of experiences, those members who met were united on one main point: that the political party which had initially emerged in answer to a security problem must stay and prosper by all possible means, for it had become the main hope for Islamic revival in the country. They decided to give priority to improving the movement's internal structures and publishing an official newspaper, and to increasing the movement's effectiveness on the national scene and its presence in the national youth, cultural and trade union organisations.

Theoretically, the movement could also have chosen to retreat from the political scene in order to adopt a more cultural and religious role, thus avoiding a repetition of the unproductive debate of 1981 concerning the "legitimacy" of an Islamic political party. However, this alternative was totally ignored in favour of making the movement bigger and stronger. The logic of the time was that everything could be achieved by capable organisation, and the tempting prospect of taking power was so great that no one opted for a change in direction. The presidential amnesty, along with all the other political indicators, was interpreted as "the green light for the movement to continue with its public activities. Thus the 1984 conference was clear in supporting an increased presence in the various national sectors and organisations."[26]

As all sectors were agreed on the political dimension of the movement's new strategy, the main disagreement was about who should lead the party: Ghannouchi or Karkar. Those who had been in prison knew the extent of the rivalry between the two men, and some of them publicly expressed their fears about Ghannouchi being re-elected to the post of *amīr*. However, many of the movement's members within the university and the local provinces regarded Ghannouchi as "a symbol" of the movement, and they ignored Karkar's objections and re-elected his rival to the position of leader. By way of consolation, Karkar was selected to lead the *majlis al-shūrā*. In doing so they tried to ensure a balance between the executive and the legislative authorities. As for Mourou, he was not considered a serious challenger to the main posts, and was confirmed as secretary-general only to benefit from his public prominence.[27]

In line with these internal and public arrangements, the party continued to play the double game of running both a secret organisation and a public political party. Accompanied by strong local and international media interest, hopes were high and expectations were great.

The new leadership chose to act on three main points. First, it tried to reorganise the central and local structures of the movement around the country, benefiting from the return of many prominent activists. Almost all local executive committees were renewed, and the main leaders were elected either to the executive bureau or to the *majlis al-shūrā*. Second, it officially announced its new public political bureau on June 6, 1985, with Ghannouchi as president, Mourou as secretary general, and Jebali as co-ordinator for political liaison with other parties.

These politicians were officially received by the Tunisian prime minister for consultation on 10 October 1985, two days after the Israeli raid on the Palestinian Liberation Organisation's headquarters in the suburbs of Tunis. The national television channel showed scenes from the meeting, which seemingly implied official recognition of the movement. The MTI was also strongly represented in the sixteenth national conference of the UGTT, still the second strongest political organisation after the ruling party, and also participated in the national conference of the Tunisian League of Human Rights, securing a place on its executive committee. Although the movement was not given permission to publish its own newspaper, its news always featured prominently in most of the local newspapers.

Thirdly, the student wing of the MTI succeeded in securing the signatures of more than 15,000 students to proceed with a general conference which was finally held from 18 to 20 May 1985, and which gave birth to the Tunisian General Union of Students (UGTE). This was seen to be an important political victory for the MTI, for it confirmed the Islamists' domination at the University of Tunis.

At this point it appeared as though all conditions for the movement were favourable, despite minor nuisances from the government, such as its refusal to grant official recognition to the MTI, or to allow certain prominent Islamists to return to their jobs in educational institutions or public administration. Ghannouchi began to show more aggression in criticising his opponents and more optimism concerning his movement's future, as we may see from the quotations below:

> The so-called liberation movements which appeared in the Maghreb to fight the colonisers took power and then betrayed those people who had fought for Islam.

> Since then they have made more efforts to paralyse Islam than the colonialists themselves.
>
> The Westernised élites presently in power in our country represent only a small minority imposed by the force of state, the army and the mass media on a population of Muslim believers. They were educated by the colonisers and from them they inherited power. The future élite which will govern Islamic Tunisia is the new generation now being persecuted. They will re-establish Islam in this land *in shā'Allāh*.[28]

Despite its apparent successes at that time, the MTI was not the only, nor was it the most important player on the Tunisian political scene. The trade unions were still the only centre of serious opposition to the social and economic policies of the régime. The economic situation that had led to the events of January 1984 remained largely unimproved, and the tension between the government and the trade unions was growing fast. What is more, the régime itself was not united owing to the deteriorating health of President Bourguiba. Although Prime Minister Mzali was the president's official successor, he was disliked by other rivals within Bourguiba's entourage, many of whom were plotting against him and working towards their own ends. Indeed the question of who was to succeed Bourguiba soon emerged as the most important issue in Tunisian political life. To all intents and purposes the president was still active on television, but the truth as seen by Tunisians was different, as one Western reporter illustrated:

> Each evening on the T.V. news, Tunisians watch a familiar sight: "President-for-life" Habib Bourguiba sits in his office at the Carthage Palace near Tunis, receiving ministers. The pictures always look the same, the narration barely changes.
>
> To many Tunisians, those frozen T.V. images illustrate their country's predicament. Tunisia today is in a period of suspended animation, waiting for a change of power.[29]

Despite strong confrontations with the UGTT and the imprisonment of its leader, Habib Ashour, the economic situation failed to improve.[30] Bourguiba's "solution" was unremarkable: in July 1986 he dismissed Prime Minister Mzali and appointed Rashid Sfar, his economic minister, in his place. Like all other Tunisians, Mzali first heard the news on television, and was left regretting the fact that he had not submitted his own resignation rather than being sacked. He later wrote: "I should have resigned, but didn't because of the state's national interests. I regret today not doing so and admit that it was a political mistake."[31]

However, the appointment of Sfar failed to stop speculation about the main issue preoccupying Tunisian politicians: Bourguiba's succession.[32] Internal conflicts between the various factions of the government intensified, and each group struggled to be in a better position to succeed the ageing leader, who by this time had begun to show agitation with both his aides and opponents. Only

Bourguiba himself was unimpressed by all this talk; he was "the only one not to think about it. He had recently announced to his party's central committee that the issue of succession was not urgent—now or tomorrow—for he would try to live many more years."[33]

Although the whole country suffered as a result of this fragile situation, the Islamists were those who most feared a renewed, direct participation by Bourguiba in the day-to-day running of the country's affairs. Bourguiba was their historical enemy, and they had heard him saying at the ruling party's conference in June 1986 that he would do his best to get rid of the shanty-towns and "*al-Ikhwānjiyya*," as he referred to the Islamists.

How best to respond to these worrying changes was the main question concerning the MTI leaders meeting at their general conference in December 1986. They agreed on a three-phase strategy which they hoped would lead them to power: first was a stage of presenting the movement and its ideas to the public, then one of preparing alternative Islamic programmes and training personnel to run an Islamic government, and finally a stage of victory and government itself.[34]

By applying this strategy there was now no way back from the political venture on which they had embarked in pursuit of their much-coveted goal: disposing of Bourguiba and taking his place in the leadership of Tunisia, rather than merely ruling their own small, secret organisation. But what if the authorities did not allow the Islamists to present themselves before the public? A military coup was suggested by the movement's secret military wing as the best solution in the event of further repression by the state. A committee headed by the military wing's leader, Mohamed Shammam, was formed and was advised to prepare plans for a possible coup in which all sections of the movement were to participate. By underscoring this final point, the civil leadership was trying to avoid total dependence on the military wing.[35]

This represented a turning-point in the Islamists' strategy. The military option was in fact contradictory to all of their official rhetoric, but for those working underground it seemed to represent the ideal means by which to conquer the Tunisian political arena and to realise their dream of an Islamic state. It was none the less also a very risky and dangerous option, because although the Tunisian régime may have been weak in the field of ideas, it was by no means weak in the field of security. By shifting to a military strategy, the Islamists were up against an experienced, tough and efficient security apparatus, led by a highly experienced and ambitious general, Zein al-Abidine ben Ali, who was then interior minister.

A few months later, on 9 March 1987, Ghannouchi was once again arrested after insisting on giving a lecture in a mosque, an action interpreted by the authorities as a political, not a religious, act. By 24 April more than 200 Islamists were behind bars and the crown prosecution ordered an official investigation, having charged them with treason, attempting to overthrow the government and defamation of the President.[36] This time it appeared that Bourguiba was determined to eradicate his enemies; he was widely reported to favour "severe verdicts to be immediately carried out—in the case of death penalties, for instance—to wipe out once and for all the fundamentalist trend."[37] This was very clear to all parties concerned, not least to Karkar, who had assumed the leadership of the MTI following Ghannouchi's arrest. He had always favoured the radical option of opposing the régime, and now found himself with an official mandate from his movement to implement his views in the face of an all-out assault from the government. Indeed, all factors were in place for what was to be the last major battle against Bourguiba.

The End of Bourguiba's Era

President Bourguiba was in his native coastal town of Monastir at the beginning of August 1987. The third day of the month was his eighty-fourth birthday. For a whole month there was to be a daily ceremony to celebrate the occasion, and each province in the country had to present the best of its singers, dancers and poets to perform in front of "*al-mujāhid al-akbar*" and his many guests at the presidential palace in Monastir. This was an annual routine invented by the state as part of the personality cult that Bourguiba had cultivated for himself and publicly enjoyed.

But that year was to follow a different course of events, when, on the night of 2 August four bombs exploded in four hotels in Monastir itself and the neighbouring town of Sousse. Thirteen people were injured, among them Western tourists from England and Italy. Bourguiba ordered his interior ministry to increase the pressure on the Islamists and arrest those responsible for the explosions. On 17 August Tunisian state television showed the video-taped confession of a Mehriz Boudagga, who, it was alleged, was the leader of the guilty group. Boudagga claimed that he had received instructions from the hierarchy of the MTI, but in three statements distributed in Paris the movement "denounced the bombings and the use of violence in general" and stated that the video shown on Tunisian state television was "the first time that they had heard of a Mehriz Boudagga, and that he was in no way connected with the MTI."[38]

The battle between the authorities and the Islamists thus reached its peak and dominated the country's political scene. The entire state apparatus was mobilised in the battle against the MTI on direct orders from the President, whereas the Islamists were well prepared in their counter-attacks: demonstrations were held, tracts were delivered, and communiqués were sent to the local and international media. The French media in particular, which had always enjoyed the special regard of the Tunisian élites, were intensely interested in the course of events, and seemed surprisingly unimpressed with the régime's claims against the Islamists, to the point that some observers suggested that France may have been implicated in some kind of conspiracy to destabilise

Tunisia, by manipulating the fundamentalists and bringing about a new, liberal and more stable régime.[39]

From the outset of the new confrontation between the régime and the MTI, the French media kept a close eye on all developments, emphasising the difficulties facing the authorities and the dangers of it going too far in fighting its enemies. *Le Monde*, for example, consistently covered the details of events from the moment of Ghannouchi's arrest. When it finally devoted its front page editorial to the aftermath of the Monastir and Sousse explosions, it did not hesitate to question the régime's accusations against the Islamists and to place the real blame on the régime, which it described in the following terms: "A régime whose main preoccupation is to not displease President Bourguiba, who, from his palace, continually manipulates his successors. The timid attempts at liberalisation made by Mr. Mzali have long been extinguished. The risk today is that the Islamists represent the only credible opposition."[40]

It was not only the French who let Bourguiba down publicly. Almost all the leaders of the opposition liberal and leftist parties stood by the Islamists in one way or another, especially the leader of the Mouvement des Democrates Sociaux (MDS), Ahmed Mistiri, an ex-minister of Bourguiba's, and his party colleagues who happened to dominate the Tunisian League for Human Rights (LTDH). This is how an angry Mohsen Toumi has described their role in the fight:

> As soon as the state responded, as forced by its duties, to the operations of destabilisation and the terrorist acts of the Islamists, by arresting those responsible and bringing them to justice, the MDS and the LTDH broke out, denounced fascism, wanted to storm the prison doors, protested against the judges and alarmed international opinion. They even went so far as to call for direct intervention by France and the United States. The French media relayed with eagerness this defence of the Islamists. Not a single day passed without Mr. Mistiri and the animators of the LTDH speaking or being quoted by the newspapers, the radio or the French television.[41]

The Tunisian authorities tried everything to justify their actions against the MTI. Two weeks after arresting Ghannouchi, it had broken off diplomatic relations with Iran, accusing its embassy in Tunis of establishing contacts with certain extremists in order to destabilise the country.[42] It was an act calculated to portray the presence of the MTI as the product of a foreign conspiracy by Islamic leaders in Tehran, and also to show the determination of the government in going as far as was necessary—as certain officials told *Le Monde*—"to put an end once and for all to the intrigues of obscurantists and propagators of sclerotic and backward doctrines," adding that President Bourguiba had taken this on "as a personal matter."[43]

This claim of an Iranian connection was restated when the trial of the 60 most prominent leaders of the MTI by the state security court began on 27 August. The prosecutor-general accused the Islamic movement of "attempting to overthrow the régime in collusion with a foreign state [Iran], holding arms, attacking the security forces, calling for insurrection, slandering the President and the government and diffusing false news and tracts."[44]

The nature of these accusations confirmed the fears of many observers both inside and outside Tunisia that Bourguiba would not be satisfied with prison sentences alone, but would seek the death penalty for the MTI's leaders. Although all the procedures of the trial were contested by the defendants' lawyers and international organisations such as Amnesty International, the government showed complete determination to end the Islamists' threat, as one observer at that time wrote:

> Tunis is starting to look more like a city that, if not yet embroiled in civil strife, is at least preparing for such. The signs of the crackdown are visible in virtually every part of the capital. Armed police patrol the street and stop citizens at random, checking their identification papers. Vans of police reinforcements, their windshields covered with wire-mesh shrapnel protection, are stationed around the city. At the University of Tunis van loads of soldiers and plain-clothes men are posted just outside the gates, and military helicopters circle overhead.[45]

During the days of the trial, there were no demonstrations held by the Islamists, although Hammadi al-Jebali, signatory to their official statements, gave the blatant warning that they would not remain calm under any circumstances:

> We will not watch this plan of extermination with our arms folded. Our supporters, who have managed to stay patient by keeping calm and counting on God's help, will not accept in any circumstances that their leadership be taken, by treachery and injustice.... If the government passes the point of no return [i.e., the death penalty], our movement, in a position of legitimate defence, will react against that tyranny.[46]

So serious was the threat of death sentences on Ghannouchi and his friends that the governments of the United States, France, Saudi Arabia and Algeria discreetly pleaded with President Bourguiba to avoid this measure.[47] Even inside the government, the strong man at the interior ministry, General Ben Ali, was known to oppose the option of capital punishment in order to deprive the MTI of unnecessary martyrs.[48] This was the same argument used in another front-page editorial in *Le Monde*: "Some of the accused have not disguised their hopes of becoming martyrs. The biggest service that the régime can offer them is to make this dream possible."[49]

Ghannouchi himself appeared ready for the worst, having given his responses to the judge's questioning. After advising his supporters not to resort to violence, he declared, "I am a human being and I want to live longer. But if it is God's will that I become the martyr of mosques, then so be it. However, I warn you that my death will not be in vain, and that from my blood Islamic flowers will flourish."[50]

It transpired, finally, that it was not to be God's will. When the verdict was announced in the very early hours of Sunday 27 September 1987, Ghannouchi did not appear on the list of the seven people who had received the death sentence. Five among them were tried *in absentia,* the other two were convicted of directly masterminding the August explosions, which, in all cases, the MTI had said that it had nothing to do with. The other accused were sentenced to various terms of imprisonment; Ghannouchi was sentenced to life imprisonment.

The verdict was regarded as lenient by Tunisian political commentators, and both President Mitterand and Prime Minister Chirac of France were reported to be happy with the "balanced verdict."[51] The French political class maintained the pressure on its historical ally, Bourguiba, while *Le Monde* continued to express its anxiety:

> Judging by the terms of an unconvincing trial, the régime has avoided making a gross mistake that would have tarnished the image it projects to the outside world of a modern and tolerant Tunisia. But the Islamic challenge is still there, undiminished. It may be weakened after around two thousand arrests, but the MTI is not fatally injured. New militants born in clandestine have taken over from the historical figures. They are younger and more resolved.
>
> The wisest method by which to prevent a terrorist drift by Islamism would be to legalise the MTI and force it to play by the rules of the institutions. The fundamentalists will then be obliged to abandon their double message, and to expose their backward ideas to a Tunisian population that remains largely attached to the modernist gains of Bourguibism.[52]

Bourguiba remained unconvinced, and his resolve against the Islamists did not diminish. Five days after the verdict was passed, he appointed his interior minister General Ben Ali as the new prime minister to succeed Rashid Sfar. His intention was to demand a more severe verdict from the court of appeal and to mobilise all the state's resources in fighting the MTI. Bourguiba ignored the advice of his friends, and failed to see the dangers of increased repression. Meanwhile those around him were perceptive to these dangers, and as his ex-Prime Minister Mzali observed, "a divided Tunisia was at the door of revolt and civil war."[53]

More importantly, Bourguiba misjudged the true personality of his new prime minister; he mistakenly thought that he was merely using a military man to crack down on his opponents, as he had used his predecessors before for various short-term aims. This time, however, he was misled; the General had already decided that this was the most appropriate opportunity to solve the ongoing problem of succession, and that he himself should be the new Tunisian president. Indeed, it took him only 36 days to execute his plan, although it later transpired that he had not been alone in plotting Bourguiba's overthrow.

Karkar, who had fled the country for Paris and then London, and Mohamed Shammam in Tunis, had also been busy in preparing their final assault on the Bourguiban régime. No one was aware of their plans at the time, for most of their public political rhetoric focused on democratic and non-violent opposition. As a result of the trials sympathy for their movement was immense, both inside and outside Tunisia, and so they decided that it was time to take a gamble.

As Shammam presided in Tunis over the special committee of top military and civil members of the movement responsible for executing the coup, Karkar remained the overall leader, or *amīr* of the MTI, from his positions in Paris and London. Between the two men, Sayyid Ferjani, a member of the Tunis committee, shuttled back and forth in an attempt to secure agreement and understanding on all the details of the operation. His final visit to London was on 30 October 1987. There he agreed with Karkar and other exiled leaders of the movement on all the details of the military coup, which had then been fixed for Sunday, 8 November, including the first statement to be read to the public and the members of the new national government that would be formed.[54]

On his return to Tunis on 6 November Ferjani met the top command committee—with the exception of Shammam, who, despite having been arrested ten days earlier, had succeeded in not giving the police any clues about his clandestine connections.

Thus the Tunisian Islamic movement was less than 48 hours away from seizing power by a military coup to be staged in the name of a national salvation project, which, it claimed, would recognise the role of other political forces. One of these Islamists states that at that time

> Tunisia did not have a choice. It had two possibilities: either a civil war, or the departure of Bourguiba. We had hoped for his eviction for a long time, but would have preferred that it had come from somewhere else. We had indirectly suggested to Mzali: Go ahead … you are responsible, you are Prime Minister … go ahead. We delayed until the last minute … because it was not just the movement that was threatened, but the entire country. It was necessary to move; but in which direction? Not by taking power—that was out of the question—but by removing Bourguiba, because he had become crazy, and by giving political power back to the country.[55]

This quotation in fact is in some ways contradictory, because "removing Bourguiba" by force would in itself mean seizing political power, even if the movement's pronounced intentions were only to govern for a short period and to organise a fair election. Another Tunisian Islamist has likened this situation to that of other groups in the Arab world who organised military coups, declaring that they wished only to serve their people and return political power to them:

> It was, in fact, a scenario in the Sudanese fashion, in the style of Siwār al-Dhahab [who, as Minister of Defence led a military coup against Numeiri in 1985 and then gave power to an elected government the following year]: to eliminate Bourguiba, but not by force. We were very clear on that point. He should not be hurt, ever. We also planned to involve everyone, all the parties, because we were aware that we were not ready to assume the heritage of Bourguiba and were incapable of solving the country's problems. So it was necessary to make everyone participate, to call on everyone to take responsibility ... and to install a government of public salvation, of national salvation, a parliamentary régime where all political parties participate in the country's evolution. It was clearly not conceivable to take power. All we sought was a democratic climate that would be more or less favourable to dialogue, to the evolution of ideas and to peaceful change.
>
> So why take power? The movement's strategy was to eliminate Bourguiba, not as a person, but as a system of decision-making.[56]

Thus it seemed that everything was in place to execute an extraordinary plan that may have brought about the Arab world's first Islamist government. One vital piece of information remained missing, however: what was the plan of General Zein al-Abidine ben Ali? It was not until half past six the following morning, Saturday, 7 November 1987, that the Islamists were to discover the prime minister's intentions which, once clarified, dealt a severe blow to their plans, calculations and ambitions.

Bourguiba was overthrown, and Ben Ali achieved all that the former prime ministers had failed to do. He deposed the ageing leader, declaring him unfit to continue as president. As expected, he also announced that he was to take over as the new president of the Republic of Tunisia. The Islamists looked set to start all over again with a revitalised Tunisian state.

Notes

1. Al-Ghannūshī, Ḥarakat al-ittijāh al-islāmī fi Tūnis, op. cit., p. 148.
2. Ibid., p. 121.
3. Anas al-Shābbī, al-Ta arruf al-dīnī fī Tūnis (Tunis: La Presse, 1991), p. 7.
4. Ibid., p. 25.
5. Muḥammad al-Hāshimī al-Ḥāmidī, Ashwāq al-hurriyya (Kuwait: Dār al-Qalam, 1989), pp. 75–6.
6. Ibid., p. 78.
7. Al-Ghannūshī, Ḥarakat al-ittijāh al-Islāmī fi Tūnis, op. cit., p. 159.
8. Al-Manṣūrī, al-Ittijāh al-islāmī wa Burqayba, op. cit., p. 10.
9. Boulby, 'The Islamic Challenge: Tunisia since Independence', op. cit., p. 609.
10. Waltz, "Islamist Appeal in Tunisia," op. cit., p. 68.
11. Al-Ghannūshī, Ḥarakat al-ittijāh al-islāmī fi Tūnis, op. cit., pp. 144–5.
12. Toumi, La Tunisie de Bourguiba à Ben Ali, op. cit., p. 223.
13. Al-Manṣūrī, al-Ittijāh al-islāmī wa Burqayba, op. cit., p. 37.
14. Al-Ghannūshī, Ḥarakat al-ittijāh al-islāmī fi Tūnis, op. cit., pp. 151–2.
15. Mohamed Mzali, Tunisie: Quel avenir? (Paris: Publisud, 1991), pp. 30–1.
16. Waltz, "Islamist Appeal in Tunisia," op. cit., p. 69.
17. Al-Ḥāmidī, Ashwāq al-hurriyya, op. cit., p. 236.

18. Ibid., p. 106.
19. Al-Manṣūrī, al-Ittijāh al-islāmī wa Burqayba, op. cit., p. 36. See also al-Ghannūshī, Ḥarakat al-ittijāh al-islāmī fi Tūnis, op. cit., p. 122.
20. Al-Ghannūshī, Ḥarakat al-ittijāh al-islāmī fi Tūnis, op. cit., pp. 154–7.
21. Quṣayy Ṣaliḥ al-Darwīsh, Yaḥduthu fī Tūnis (Paris: Q. Darwish, 1987), pp. 168–9.
22. Rashid al-Ghannūshī, interview, Arabia, April 1985.
23. 'Abd al-Fattāh Mūrū, interview, Al-Mujtama', 15 January 1985.
24. Al-Ghannūshī, Ḥarakat al-ittijāh al-islāmī fī Tūnis, op. cit., pp. 105–6.
25. Al-Manṣūrī, al-Ittijāh al-islāmī wa Burqayba, op. cit., p. 81.
26. Ibid. p. 41.
27. Waltz, "Islamist Appeal in Tunisia," op. cit., p. 71.
28. Rāshid al-Ghannūshī, interview, Arabia, April 1985.
29. Louise Lief, "Tunisia Wary of Future Leadership Void," Christian Science Monitor, quoted in The Movement of Islamic Tendency in Tunisia: The Facts, op. cit., p. 57.
30. Alan Cowell, "Tunisia Uneasy about Future as Bourguiba Ages and his Grip Tightens," New York Times, 25 June 1987.
31. Mohamed Mzali, Lettre ouverte à Bourguiba (Paris: Editions Alain Moreau, 1987), p. 29.
32. Lief, "Tunisia Wary of Future Leadership Void," op. cit., p. 58.
33. Al-Darwīsh, Yaḥduthu fī Tūnis, op. cit., p. 199.
34. Al-Manṣūrī, al-Ittijāh al-islāmī wa Burqayba, op. cit., p. 63.
35. 'Abd Allāh 'Imāmī, Taẓīmāt al-irhāb fī al-'ālam al-islāmī: unmūdhaj al-Nahḍa: al-nash'a, al-taẓīr, al-haykala, al-irhāb ... (Tunis: Al-Dār al-Tūnisī li-l-nashr, 1992), pp. 256–7.
36. The Movement of Islamic Tendency in Tunisia: The Facts, p. 18.
37. Toumi, La Tunisie de Bourguiba à Ben Ali, op. cit, p. 205.
38. The Movement of Islamic Tendency in Tunisia: The Facts, p. 20.
39. Toumi, La Tunisie de Bourguiba à Ben Ali, op. cit, p. 220–2
40. Le Monde, 6 August 1987.
41. Toumi, La Tunisie de Bourguiba à Ben Ali, op. cit, p. 219.
42. Le Monde, 28 March 1987.
43. Ibid.
44. Le Figaro, 28 August 1987.
45. The Movement of Islamic Tendency in Tunisia: The Facts, p. 16.
46. Hammadi al-Jebali in Le Monde, 11 September 1987.
47. Le Point, 21 September 1987.
48. Toumi, La Tunisie de Bourguiba à Ben Ali, op. cit, p. 205.
49. Le Monde, 16 September 1987.
50. Ibid.
51. Le Figaro, 29 September 1987.
52. Le Monde, 29 September 1987.
53. Mzali, Tunisie: Quel avenir?, op. cit., p. 21.
54. 'Imāmī, Taẓīmāt al-irhāb fī al-'ālam al-islāmī, op. cit., pp. 263–75.
55. Francois Burgat and William Dowell, The Islamic Movement In North Africa (Austin: University of Texas Press, 1993), pp. 226–7.
56. Ibid.

4

Islamists v. Ben Ali: 1987–1993

By the early morning of 7 November 1987, Tunisian radio had broadcast the communiqué that would bring an end to Bourguiba's presidency. Its author was his former prime minister, General Zein al-Abidine ben Ali. The statement was brief and succinct: it began by praising Bourguiba's role in the liberation and development of Tunisia, then gave the reasons for his eviction:

> The onset of his senility and the deterioration of his health, and the medical report made regarding this called us to carry out our national duty and declare him totally incapable of undertaking the tasks of President of the Republic.
>
> Thereby, acting under Article 57 of the Constitution, with the help of God, we take up the Presidency of the Republic and the high command of our armed forces.
>
> In the exercise of our responsibilities, we are counting on all the children of our dear country to work together in an atmosphere of confidence, security and serenity, from which all hatred and rancour will be banished.
>
> The independence of our country, our territorial integrity, the invulnerability of our fatherland and our people's progress are a matter of concern for all citizens. Love of one's country, devotion to its safety, and commitment to its growth are the sacred duties of all Tunisians.[1]

The new president then went on to promise a new era of democracy for the people of Tunisia:

> Our people have reached a degree of responsibility and maturity where every individual and group is in a position to constructively contribute to the running of its affairs, in conformity with the republican idea, which gives institutions their full scope and guarantees the conditions for a responsible democracy, fully respecting the sovereignty of the people as written in the Constitution. This Constitution needs urgent revision. The times in which we live can no longer accept a life presidency or automatic succession, from which the people are excluded. Our people deserve an advanced and institutionalised political life, truly based on the plurality of parties and mass organisations.

> We shall soon be putting forward a bill concerning political parties, and another concerning the press, which will ensure wider participation in the building up of Tunisia and the strengthening of her independence in a context of order and discipline.
>
> We shall see that the law is correctly enforced in such a way that will proscribe any kind of iniquity or injustice. We shall act to restore the prestige of the State and to put an end to chaos and laxity. There will be no more favouritism or indifference where the squandering of the country's wealth is concerned.[2]

Islam was also given mention in the statement, but not in the context that the Islamists had hoped for. It was referred to in the international obligations of the new government:

> We shall continue to maintain good relations and positive co-operation with all other countries, particularly friendly and sister countries. We shall respect our international engagements.
>
> We shall give Islamic, Arab, African and Mediterranean solidarity its due importance.
>
> We shall strive to achieve the unity, based on our common interests, of the Greater Maghreb.[3]

Thus Tunisia found itself with a new president, with no hint of opposition or disapproval. Evidence of the excellent preparation and execution of the plan emerged when it was revealed that Bourguiba had failed to command any kind of support to prevent his eviction from Carthage Palace, and that no one had been killed during his arrest.

For the supporters of the MTI it was a day to be celebrated, in that the turn of events could potentially save the heads of their leaders. But for these leaders themselves, both in Tunis and overseas, it was a grave moment, for they had been "robbed" of a historical coup that was to have been carried out less than 24 hours later.

Only they and the new president were aware of this. Ben Ali, who had been preparing his move for weeks, had discovered his rivals' plot just three to four days before their plan was to have been carried out, after having arrested a soldier who had been invited by the Islamists to participate in the coup. Although the soldier had not provided a complete account of their plan, the details were sufficient for Ben Ali to decide to act early and seize power before Karkar and his colleagues could do so.

The question remaining for the Islamists was problematic: should they proceed with their plan, as was technically still possible, or should they abandon it, since it was becoming clear that the new régime was on full alert in anticipation of rebellion? Karkar was insistent that the Islamists should stick by their first option, while the majority of Islamist leaders in Tunis deemed it impossible to proceed with their original plan and agreed to call it off. They succeeded in winning the argument, because it was they who directly controlled the military personnel who had been preparing for the coup, most of whom knew that they would be the immediate targets of Ben Ali's régime. A few days later many of them were arrested, including Sayyid Ferjani who, after being tortured, revealed to the authorities new details of the aborted coup.

This was the logistical argument behind ceasing the movement's military action. The political argument was even stronger, because the overthrow of the president had been widely welcomed both inside and outside Tunisia, almost unanimously so. Even for Mohamed Mzali who, on publishing his future vision for Tunisia in 1991, vigorously condemning that of Ben Ali, asserted that the events of November 1987 were a time of enormous popular relief:

> People did not closely question the way in which it had been accomplished. They did not seem preoccupied by the fact that the operation had been brought about by a palace revolt, between the men of the seraglio, that the people took no part in it, that the Prime Minister, the ex-Interior Minister of the fallen President was the main actor, and that the seven doctors summoned to the interior ministry at two o'clock in the morning to write a certificate of ineptitude did not examine "the patient," nor did they determine when the illness had started.
>
> The relief was so real that these legal preoccupations were seen as pedantic quibbles. The worst had been avoided, and this was what mattered most. In addition to that, the self-proclaimed successor to President Bourguiba had the ability to present himself, in his "declaration" of 7 November, as a reformist anxious to turn the page of the past in a convenient manner, and to engage Tunisia on the path of democracy and development. The new president found himself at one of those cross-roads that history offers in the course of a nation's life to make possible the realisation of its main aspirations. General ben Ali benefited from a quasi-general consensus, both inside and outside the country.[4]

This was the kind of acceptance and hope with which Tunisians, the majority of Islamists included, received their new president. Ben Ali found no trouble in securing his grip on both the state apparatus and the ruling party, for he had been the key player in the country for the last few years of the Bourguiba era. As recorded in his first declaration, he did not present himself as the leader of a radical revolution that aimed to introduce widespread change, rather he preferred to be the "saviour" of the republic, a role that had hitherto been played by Bourguiba.

The Strategy of the New President

It was within this context that he chose not to dismantle the long-standing Constitutional Socialist party, through which Bourguiba had ruled Tunisia since 1956 in defiance of some observers who believed that he should do so. Rather, he reinforced it by becoming its new president, and appointing his Prime Minister Hedi Baccouche as its secretary-general. The only notable change introduced was in the party's name, which on 27 February 1988 became the *Rassemblement Constitutionnel Democratique* (RCD).[5]

A new law for political parties was announced on 3 May that same year, and a law for the press was announced on 2 August. Although both were slightly different from previous laws, they failed to have any significant effect on the political climate, for the first law insisted once more that no political party could be based on either religious values or Arabism, thus barring the way for the political ambitions of the Islamists as well as the less influential Arab nationalists.[6] The mood of the new president was cautious; he was determined to do things his way under his control, which was very much the way in which he dealt with the Islamists.

Viewed from the perspective of Bourguiba's intentions to eradicate the Islamists, there was a tremendous sense of relief following his fall. The new president, however, did not want his success to be judged according to the MTI's popularity. He followed closely police interrogations of the members of what was later called "the security group," meaning those who were implicated in the Islamists' aborted coup, and was aware of just how close they had been to taking his place. Nevertheless, he adopted a well-measured plan with which to diffuse the tensions with the Islamic movement, without giving in to their main demands.

First, the police crackdown on civilian members of the movement was halted, but those already in prison had to wait a few months before receiving a presidential amnesty. When Ali Laridh, one of the MTI leaders who had been sentenced to death *in absentia* was arrested, his sentence was confirmed by a court of the new era and he too had to wait for a presidential amnesty. Ghannouchi had to wait until 1 May 1988 before finally being granted amnesty and released. He emerged from prison highly optimistic, announcing to journalists that he had confidence in God, then in Ben Ali, to make a better future for Tunisia. For him, this was a form of conditional *bay'a*: "The current ruler of Tunisia made a coup against Bourguiba promising two things: Bourguiba failed to implement democracy but I will, and Bourguiba disregarded the Arab and Islamic identity of Tunisia, which I will restore. On these bases we gave him *bay'a*, as did the people."[7]

By June 1988, Ben Ali announced that one of his major priorities was to find a happy and final solution to the Islamist problem, and allowed the return from exile of Abd al-Fattah Mourou, the secretary-general of the MTI, and Hammadi Jebali, its former leader. They and others who emerged from hiding had to stand in court once more in October 1988. They were again found guilty of being members of an illegal organisation and sentenced to five years in gaol, then were allowed to go home from the court, in accordance with higher instructions, and were later amnestied on the first anniversary of the events of 7 November.

There were other positive signs: a member of the MTI was included in a national committee that had been asked to draw up a "national pact" on the platform of a democratic system; the student organisation (UGTE), formed by the Islamists in 1985, was given official recognition; and Ghannouchi was invited to hold talks with a number of the president's top aides, to whom he was to put his demands and sort out the problems of those returning from exile or coming out of gaol and reclaiming the jobs they had lost during the confrontation with Bourguiba.

One of the most sensitive issues was resolved at this forum: Ghannouchi agreed to distance publicly his movement from the members of the "security group" in return for their quiet release from gaol. Although it was not an easy decision for the Islamists' leader to take, he was presented with no other choice if he was to reassure the public and President Ben Ali that his movement was truly civilian. For the new régime, it was also a test to see how far the Islamic-political stand could adapt to a stable political climate, and was of course an obligation if the declaration of 7 November was to have any credibility—at least for its first two years. As Abdallah Imami, a defender of the government argues, the new régime predicted two responses from the Islamists to its policy of appeasement:

> [First] the movement will reform itself and play by the rules of legitimacy and law, which implies a break with the legacy of secrecy and the adoption of one field of action only: public life, a choice which makes it compulsory to dismantle the military and intelligence wings and the whole secret apparatus of the movement.
>
> [Second] the movement will maintain its characteristics, thus limiting its public presence to the activities of certain known figures, while its main strength will be concentrated in the secret organisation, which will make only a few necessary adjustments—mainly to replace those uncovered by the police with new faces and clean records.
>
> On the basis of this analysis, the government defined its responsibility to take a soft line that would make it easy for the movement to proceed towards legalisation.[8]

Parallel to this soft line taken against the MTI, the new régime adopted what looked like a more positive stand towards Islam in general, as part of the promises made in the new régime's first statement. Three weeks after taking power, Ben Ali took the decision to broadcast on radio and television the call for the five daily prayers.[9] The sighting of the moon was accepted as a method for deciding the days of the two main religious festivals of the year, reversing Bourguiba's previous directives to use the Christian calendar. On the last day of 1987, the president announced the reopening of the Zeitouna University to accommodate three colleges of religious studies. An Islamic High Council formed in April 1987 was reactivated by a new presidential decree, and Mourou, the secretary-general of the MTI, was invited onto its membership.

Another similar decree passed on 30 March 1990 announced the formation of a centre for Islamic studies in Kairouan, while a new ministry for religious affairs was created and attached to the prime minister's office, later becoming an independent ministry. Two annual presidential prizes for memorising the Qur'an and for Islamic studies were also announced.[10]

One other main act of legislation introduced was also very relevant to these new measures: a new law for mosques was announced on 3 May 1988, which made it illegal for anyone to give lectures there without prior authorisation from the prime minister, thus making it very difficult for the Islamists to continue to propagate their views in their traditional forum.[11]

Within the freer climate of expression that prevailed throughout that period, every positive or negative step taken in the Islamic field was closely monitored by all political groups in the country, not least by those considered leftists and secularists:

> The secular opposition parties, who feared the fallout that could result from an attempt to pull the rug out from under the feet of the Islamists, had undertaken to mobilise themselves and to demand that the régime states clearly the limits of the recentering of its position on religion. On 18 March, forty university professors representing the principal political formations published a long declaration calling for the necessary separation of Islam and politics, and the obligation to preserve vigorously the acquisitions relating to the personal status code.
>
> On March 19, on the eve of the new national holiday, President Ben Ali put an end to the different interpretations of his cultural policy during a televised speech, defining for the first time the limits within which he intended to confine his action from then on: "There will be no calling into question, nor abandoning of that which Tunisia has been able to achieve to the profit of women and the family. The personal status code is an attainment to which we are attached and by which we feel attached. We are proud of it, and we draw real pride from it."[12]

What is more, the question of whether to allow the MTI to operate as a legal political party remained unclear. Although the movement signed the new National Pact with the government and the other main opposition parties on the first anniversary of Bourguiba's overthrow, this was not enough to solve the problem of official recognition. The new Parties Law unequivocally banned religious parties, thus obliging the Islamists to change their name for the second time and to make more concessions in order to be accepted within the new system.

By the end of 1988 the MTI had reached a decision to change its name to *Ḥarakat al-Nahḍa*, or the Renaissance Movement. Internal discussions regarding this point, among others, were problematic because of the presence of a number of strong personalities within the movement's leadership, and also because the structure of the movement continued to be based on a secret apparatus controlling the public presence.

Ghannouchi was announced as president of this new political party, but its real president was not known publicly until his arrest two years later. This was Dr Sadiq Shourou, a Professor at the University of Tunis, elected to the post at a secret general conference held by the MTI in March 1988, less than two months before Ghannouchi's release from prison.[13] The third most powerful man in the movement was again Mohamed Shammam, who regained his control over the security wing of the movement. Karkar remained in France and rejected the idea of returning to Tunis for he claimed to have no confidence in the new régime.

The new party's manifesto was not very different from that drawn up by the MTI in 1981, although there was a clear effort in it to limit the use of the word "Islam" as much as possible, so as to avoid breaching the rules of the new Parties Law. Its primary objective was the preservation of the republican régime and its bases, the preservation of civil society (a concept absorbed from the secular literature of that time), and the implementation of the principle of peoples' sovereignty and the realisation of *shūrā*.[14]

In February 1989, a new official application was made to the interior ministry to grant legal recognition to the newly named party. The interior minister was obliged by law to provide an answer, either positive or negative, within four months. This proved to be a very long and significant period in the fragile relationship between the Islamists and the régime, mainly as a result of the general elections which were due to be held on 2 April that year.

Ben Ali Against the Islamists

When the application for the new political party was submitted, hopes were high among the Islamists that a positive response awaited them this time. The public wing of *al-Nahḍa*'s leadership spared no efforts in allaying Ben Ali's fears about the movement's intentions. The unspoken understanding was that in return for official recognition, *al-Nahḍa* would avoid any form of challenge for political power, for it regarded itself primarily as a movement of *da'wa* and social change. This understanding was made clear by the movement's main leaders, especially by Mourou to Salah Jourshi, the former member of the movement who had mediated between the MTI and the government during the first half of 1988.[15]

In line with this understanding, the movement approved the idea suggested later by the ruling party that all those who had signed the National Pact should present a common list of candidates for the general elections. However, Ahmed Mistiri's MDS rejected the idea shortly before the elections, making the hope of any other kind of coalition problematic for the Islamists, for the MDS was still deemed the most important opposition party.

The movement's first reaction was to present 5 candidates for 141 contested seats in 25 constituencies, representing a symbolic presence with which to gain a few more supporters without alarming Ben Ali or any of the other parties. This argument was later overruled by those of the radical faction who controlled the movement secretly, and who argued that it was the MTI that had brought Bourguiba down, and that the time for cautious calculations was over. *Al-Nahḍa* was capable—in their view—of attaining a parliamentary majority, a fact that Ben Ali and others should learn to accept and live with.

The debate escalated into what amounted to a political coup within *al-Nahḍa*, as Jourshi later recalled:

> Here came a sudden and serious change within *al-Nahḍa*'s policy. After it had accepted the principle of not participating in the elections, and maybe also the support of the RCD's candidates in return for official recognition, it suddenly decided to participate under the banner of the independent lists. Then it jumped from [the notion of] a symbolic participation in a few constituencies to presenting candidates in all constituencies, thus becoming the only challenger locally and nationally to the ruling party.
>
> Instead of opting for a symbolic presence in these constituencies, the movement's masses did their best to achieve full mobilisation and guarantee the support of thousands of members and supporters, to the point of raising the authorities' fears, who found themselves obliged to mobilise all their efforts and capacities in order to face up to this sudden change in the movement's policy.[16]

Electoral rhetoric was to prove another problem for the Islamist leadership, because the views of many of the independent candidates chosen to represent the movement did not always conform to those of Ghannouchi and his followers. As for the supporters of the Islamic movement in general, they were not especially interested in the detailed promises of the candidates, but rather in a serious Islamic alternative to 30 years of one-party rule.

The results of the election did not reflect the trends revealed in the electoral campaign, although they confirmed the worrying weakness of the secular opposition parties, who all together failed to gain 5 percent of the vote. The ruling party was declared victorious in all constituencies, and Ben Ali received 99.27 percent of the votes in the presidential elections held on the same day. He was the sole candidate.

As for the Islamists, it was announced that their average share of the vote was around 17 percent and was as high as 14.21 percent in certain constituencies,[17] the highest number of votes being gained in the capital and a number of other big cities. However, for all these votes, they failed to win any seats. As a result, they protested, and accused the government of rigging the elections. Spurred on by the desire for power, the Islamists were derisive of the government's admission of their strong showing in the elections and lashed back, as we see from this account:

> Despite the fact that the officially declared results showed the Islamists as the second political force in the country and admitted that their [total] share of the vote

> was seventeen percent, the most violent statement against the régime after the results were announced came from *al-Nahḍa*. This led to the breaking off of relations and the accumulation of reasons for tension and confrontation.
>
> *Al-Nahḍa*'s leadership ... became tempted by the electoral game and quickly jumped in such a manner that raised the fears of the government and the political élite. It suddenly changed from a political group seeking legal permission and a few seats in parliament to playing the role of a major influential force, not content to merely lead the opposition, but believing in its ability and legitimacy to claim a parliamentary majority, and [claiming] that it played a decisive role in the changes of 7 November.[18]

The April elections were yet another decisive factor as far as relations between Ben Ali and *al-Nahḍa* were concerned. For the president, the Islamists had failed his test and had shown that they were keen to enforce their political domination by every means, and were therefore not to be trusted. For the Islamists, the radical elements found new ammunition in calling for a strategy with which to counter Ben Ali's plans and relaunch the initiative that had almost brought them to power in 1987. As the cover story of the 17 April 1989 issue of the French magazine *Jeune Afrique* declared, the main result of the elections was "Ben Ali face aux Islamistes."

Confrontation between the government and the Islamists did not take long to materialise. First, *al-Nahḍa*'s application for official recognition was rejected on 8 June 1989 on the grounds that some of its founders had "criminal" records. Another application was made on 11 December that same year but remained unanswered. In a new government formed after the elections, a prominent leftist personality opposed to the Islamists was given the post of education minister, and entrusted to master a new educational vision opposed to that of the Islamic movement. On both 25 July Tunisia's Republican Day, and on the second anniversary of his coming to power, Ben Ali used his speeches to clarify his position: "We say to those who mix religion with politics that there is no way of allowing them to form a political party."[19]

This latest development created new problems among other opposition parties and raised concern about a return to repression. As a result, all of the opposition parties decided to boycott the local elections held on 10 June 1990. The MDS spoke of a political impasse because of the incompatibility between the democratic rhetoric of the government and its contradictory deeds.[20] Earlier in Paris, the Tunisian embassy had refused a request from Ghannouchi—who had left the country in May 1989—to renew his passport.

The only positive step taken in 1990 was the permission given on 8 January for the movement to publish its own newspaper, *al-Fajr*. In reality it was little more than a trap in which to gain any new, written evidence that might be used against *al-Nahḍa* in any forthcoming confrontation. The prime minister at that time, Hamid al-Karwi, said, "*Al-Nahḍa* speaks more than one language and has more than one face. We gave them permission to found a newspaper to discover their views and programmes."[21] After only a few months, the director of *al-Fajr* was accused of publishing defamatory articles, and was arrested and jailed. The newspaper itself was closed down the following year, in January 1991, just one month after a new extensive crackdown was launched against *al-Nahḍa*'s leaders and supporters around the country.

This time things were totally different from what they had been in 1987. Ben Ali was not Bourguiba: he was much younger and totally in control of every aspect of the country's political life. Nevertheless, *al-Nahḍa*'s leaders were feeling remarkably self-confident, and adopted a new strategy which they called "the enforcement of liberties." Here is Ghannouchi's definition of this strategy:

> We did not mean to seize power, as was claimed, but to pressurise the régime to the point of accepting the people's decision. Nothing in our statements—if interpreted fairly—shows that we called for violent means by which to topple the government. The central key to our strategy was the enforcement of liberties, even if this led to the fall of the régime. Our goal was not to bring down the government but to pressurise it into negotiating with the people.[22]

Logically, however, one cannot truly conceive of "enforcing liberties" on a régime if one is not strong enough to do so, and there is no doubt that *al-Nahḍa* was depending on the support it had been given in the 1989 elections, and on its continuing strong presence in the University of Tunis; that is, of course, in addition to whatever presence it still had within the army and security services. Two other external factors also raised the expectations of the Islamists: one was the overwhelming victory of the Algerian Islamic Salvation Front (FIS) in the local elections of 12 June 1990, which represented a big moral boost for Islamists all around the world; the other was the second Gulf War after the Iraqi invasion of Kuwait on 2 August 1990. The Iraqi President Saddam Hussein had gained wide support among the Tunisian people, and was seen as a brave leader who had stood up to Israel and the US-led Western alliance. Ghannouchi believed that the revolutionary mood created as a result of Iraq's policies would help the movement's plan to take the Tunisian people onto the streets and force the régime to respect their will. He gave his unconditional support to Saddam and tried to present the Iraqi president's stance as a catalyst for a global Islamic revolution.

In a passionate and long appeal he addressed the scholars and the masses of the Islamic nations, asking them to give up their lives and belongings for the leader who had dared to attack Israel, and who had stood up to the invading international forces. The conclusion of this speech, however, may be interpreted as applying primarily to Tunisia:

> Most important for you, the youth of Islam, the hope for its renaissance, the energy for its revolution, the support for its scholars ... is not to fear the great sacrifices

> required, and not to cease your blows to the heads of dictators and the invading *kuffār* and international agents, until they and their intellectual leftovers and hypocritical agents are expelled.... Fill the streets against them and their agents with your screams and protests, and destroy their foundations in your hearts and societies.... Do not allow the movement on the streets to stop, or the strikes, the confrontations, and the threatening of their interests. The Merciful has offered you a great gift through Iraq, so pick up the lead of revolution and don't let it slip away.[23]

It was during that same period that Ghannouchi recorded an audio message to the youth of *al-Nahḍa* inside Tunisia, calling on them to stand up to the government and be ready to make sacrifices for a promised victory. However, the cassette found its way into the hands of the authorities, to be used as evidence against the Islamists.

What the Islamists did not anticipate, however, was Ben Ali's clever handling of the Gulf War, in that he succeeded in using the conflict to secure fresh popular support and to rectify his problems with the other opposition parties. In the meantime, he also mastered security plans with which to deal with the demonstrations that had begun to take place, first organised by the Islamists in September 1990. Although some acts of violence occurred during confrontations between the security forces and the Islamists, nothing worked according to *al-Nahḍa*'s plans and calculations.

By December 1990, more than a hundred Islamists had been arrested, among them several members of the "security group." They were one of "three dangerous groups that were planning to execute terrorist activities," as was announced by the interior minister, who also accused *al-Nahḍa* of being implicated with these groups, and said that they would be dealt with firmly in a manner that would "uncover their hypocritical religious claims and stop them from harming the country and the people."[24]

The Panic of the Islamists

That was the official start of the confrontation. In the panic of losing their main leaders, a number of angry Islamists attacked the local bureau of the ruling party in Bab Souika, a district of Tunis, on 17 February 1991. In trying to burn down the building, they injured two security guards, one of whom died and the other was seriously burned. Ten days later, most of those who took part in the attack were arrested and their confessions were made public on television on 1 and 3 March. *Al-Nahḍa* was clearly losing the battle as far as public opinion was concerned; all the opposition parties condemned the attack and the violent behaviour of the Islamic movement, but the most outspoken condemnation came from within the movement itself.

The interior ministry invited to a meeting those few remaining leaders of *al-Nahḍa* who were not in prison, and put to them all the details gathered from the detainees implicated in the Bab Souika attack. Although the leaders condemned the attack and all other acts of violence, this was not deemed to be sufficient. On 7 March 1991, Mourou, the secretary- general, Beldi, the ex-*amīr* and current president of the *majlis al-shūrā*, and Ben Issa Demni, editor of *al-Fajr* newspaper, published a joint communiqué in which they condemned the "irresponsible acts" of violence committed with the "approval of certain leaders of our movement," and announced "the freezing of our membership and activities with *al-Nahḍa* movement, while waiting to take other steps in the near future."[25]

With this statement, the two main founders of the Tunisian Islamist movement went their different ways publicly, and their movement became divided in the midst of the most difficult period of its history. Mourou and his friends went on to announce their intention to form a new political party, to which Ghannouchi would not be admitted as a member, because, according to Mourou, "he had always objected to dialogue. He chose to resort to violence. But other Islamists want to hold dialogue with the government in office; myself for example."[26]

The government, however, expressed little interest. For Ben Ali, the priority was to suppress any opportunity for *al-Nahḍa* to make a surprise coup. On 29 March, the head office of the pro-Islamist student organisation (UGTE) was raided and was said to contain "chemical products intended for the fabrication of Molotov cocktails" and "tracts hostile to the government,"[27] the official reason given for the banning of the organisation. Violence erupted again in the university, leading to the deaths of three students during confrontations with the security forces. Instead of the support they had shown in 1987, the opposition parties signed a joint statement in which they "condemned *al-Nahḍa*'s decision to resort to confrontation" and held it responsible for putting at risk "the main national interests of the country and its security and stability."[28] The government even went so far as to invite the leaders of the opposition parties to inform them about their findings on the questioning of the arrested Islamists, before making them public.

The political isolation of *al-Nahḍa* was then all but complete, both inside and outside Tunisia. On 22 May the Interior Minister Abdallah Kallal held a press conference at which he officially accused *al-Nahḍa* of conspiring to overthrow the régime by violence. He spoke of a five-phase plan, which had begun with the distribution of tracts and the holding of demonstrations, and which had culminated in using the army, through the military supporters of the movement, to bring down the government. He confirmed the arrest of 300 Islamists, including 100 from the army, and announced that Ghannouchi and Shammam had been the masters of this plan from their chosen exile in Algiers. Journalists listened to a recorded message from Ghannouchi, in which he vowed, "the struggle will not stop before total victory, no matter how many sacrifices and martyrs."[29]

What followed later was a total attack on *al-Nahḍa* and everything connected with it, in almost every political and social aspect. Even the option of dealing with Mourou was discarded, for it was seen as a clever manoeuvre by which to preserve the Islamic movement, but in another form. Mourou was ordered to abstain from any political activities. On 28 September 1991 the government revealed the discovery of a new plot by the Islamists to assassinate the president himself, by using a "stinger missile" to hit the president's plane. It was the "top degree of treachery," according to the ruling party's official newspaper. It said in its editorial: "The obscurantists' plot to assassinate the president has closed the door to any kind of mercy towards them in the hearts of Tunisians. We now consider them—after what their sick minds have planned and their criminal hands were going to do—traitors to Tunisia and enemies to its people."[30]

Al-Nahḍa was clearly helpless; the strategy of "enforcing liberties" had ended in total disaster. Almost everything it had achieved during the last two decades had been lost, for the time being at least. The thousands of its leaders and members arrested were put on trial and given various sentences ranging from the death penalty for those implicated in the Bab Souika attack, to life sentences for most of the political leaders and shorter terms for others.

The government took its fight abroad to make it difficult for Ghannouchi to make any negative publicity against it. Although the leader of *al-Nahḍa* obtained political asylum in Britain, his field of action was much limited by the refusal of most of the Arab and international media to report his views and comments, under pressure from the Tunisian régime, and also by the continuation of differences among his colleagues. Some of them held him responsible for the confrontation, and demanded not only that he resign from the leadership, but also that he not be allowed to hold any influential position in the movement again.[31]

By the end of 1993, Ghannouchi was speaking of a "new Bosnia taking place in Tunisia without being noticed," and claimed that the government was planning to "wind up the heritage of Arab-Islamic culture in all its aspects," although he indicated that his movement "had made good progress in absorbing the régime's attack. We are now proceeding to prepare an overall plan to liberate our people from dictatorship."[32]

A different assessment of the situation was given by President Ben Ali to the French newspaper *Le Figaro*. He said:

> Fundamentalism—or if you like the religious obscurantist trend—results in terrorism.... It is a reactionary ideology that rejects all democratic values.... The fundamentalists claim they are the trustees of religion while religion is for all people. Religion is a personal matter and not an extreme closed ideology.... In my opinion, there is not much difference between what you call "moderates" and "extremists." Their final goal is the same, to form a theocratic and authoritarian state. To combat those who use religion in order to gain power by force, those who reject democracy and commit violence, we have adopted a policy of various dimensions. It starts in school and continues with wide-ranging policies in the economic and social sectors.... I do not fear Algerian contamination, Tunisia is a safe body; this

> problem is now solved. Fundamentalism is now your problem—I mean that of Paris, London and Washington.[33]

This last assessment was indeed close to reality as far as Tunisian Islamism in the mid-nineties was concerned; however, it did not mean that the fight between both men and parties was conclusively over.

Notes

1. Excerpt taken from the text of the official *Declaration of November 7th, 1987*, issued in the name of Zine al-Abidine Ben Ali.
2. Ibid.
3. Ibid.
4. Mzali, *Tunisie: Quel avenir?*, op. cit., pp. 21–2.
5. Toumi, *La Tunisie de Bourguiba à Ben Ali*, op. cit., p. 238.
6. See *November 7: al-Thawra al-hādi'a* (Tunis: A. Ben Abdallah for Publishing and Distribution, 1992), p. 176.
7. Al-Darwīsh, *Yaḥduthu fī Tūnis*, op. cit., pp. 100–1.
8. 'Imāmī, *Taẓīmāt al-irhāb fī al-'ālam al-islāmī*, op. cit., p. 286.
9. *November 7: al-Thawra al-hādi'a*, p. 402.
10. Ibid., pp. 402–21.
11. Jallūl al-Jarbī, *al-Huwwiyya fī Tūnis al-'ahd al-jadīd*, Tūnis 'ahd al-iṣlāḥāt al-kubrā (Tunis: Al-Wikāla al-Tūnisiyya li-l-ittiṣāl al-khārijī, 1992), p. 46.
12. Burgat and Dowell, *The Islamic Movement in North* Africa, op. cit., pp. 232–3.
13. *Réalités* magazine, Tunis, 24 May 1991.
14. Rāshid al-Ghannūshī, *al-Ḥurriyyāt al-'āmma fī al-dawla al-islāmiyya* (Beirut: Markaz dirāsāt al-waḥda al-'arabiyya, 1993), p. 339.
15. Salāh al-Jūrshī in *Al-Majalla* magazine, London, 13 May 1992.
16. Ibid.
17. Toumi, *La Tunisie de Bourguiba à Ben Ali*, op. cit., p. 278.
18. Al-Jūrshī, in *Al-Majalla*, London, 13 May 1992, op. cit.
19. *Le Maghreb* magazine, Tunis, 10 November 1989.
20. *Réalités* magazine, Tunis, 26 June 1990.
21. *Réalités* magazine, Tunis, 1 December 1989.
22. Al-Darwīsh, *Yaḥduthu fī Tūnis*, op. cit., p. 184.
23. This quote has been taken from an unpublished address ... *Ilā 'ulamā' al-umma wa jamāhīrihā* (London, 1991), p.11.
24. *Le Maghreb*, Tunis, 7 December 1990.
25. *Jeune Afrique*, Paris, 27 March 1991.
26. *Jeune Afrique*, Paris, 12 June 1991.
27. *Réalités*, Tunis, 17 May 1991.
28. *Al-Sabah*, Tunis, 27 May 1991.
29. *Al-Hurria*, Tunis, 29 September 1991.
30. Ibid.
31. A statement by Lazhar 'Ab'ab, published in *al-Ḥayāt*, 3 August 1994.
32. Rāshid al-Ghannūshī, interview, *Al-Da'wa*, Islamabad, 16 December 1993.
33. Zin al-Abidine Ben Ali, *Le Figaro*, 2 August 1994.

5

The Basis for a "Political" Islam

In order to analyse and understand the main ideas of the Tunisian Islamists, there are certain key concepts which must first be addressed, for they play a fundamental role in defining the Islamists' contemporary mission. The first is their religious judgment of contemporary Muslim societies, and whether these societies may be considered truly "Islamic" or not. The answer to this question paves the way for an equally important second issue concerning the difference between the "Muslim" and the "Islamist." Thirdly and finally, there is the issue of the Islamists' stand regarding the growing secularist trend in Tunisia and the Arab world, and whether or not it can be tolerated within a modern Islamic state.

The Party of Islam

The Islamists' judgment of the *religious status* of their societies both embodies the essence of their mission and highlights their main objective. Accordingly, it puts them on a different course from that of other political groups in their countries. Since their starting-point is to re-establish the authority of Islam in all areas of life, most Islamists tend both to associate themselves with Islam and dissociate others from it. It has become a priority for every Islamic movement the world over, and for all of their members, to feel, believe and show others that *they* are the representatives of the "true" Islam.

For those of the main Muslim Brotherhood (or *Ikhwān*) trend, to which *al-Nahḍa* belongs, this has been their attitude from the outset. In the first paragraph of the famous *Risālat al-taʿalīm* [Treatise of Teachings], which summarised the main points of the movement's ideology, its founder Ḥasan al-Bannā called upon every Muslim Brother to "believe strongly that our ideology is truly Islamic and to understand Islam in the way we understand it."[1] One of the movement's most prominent Syrian leaders, Saʿīd Ḥawwā, wrote on the fiftieth anniversary of the movement's creation, "the greatest way in which the party of Allah is represented in our time and region is the Muslim Brotherhood, in the way Professor Ḥasan al-Bannā shaped it."[2]

The problem, however, is that more than one group claim themselves to be the "party of Allah," and in arguing for the case of the *Ikhwān*, Ḥawwā succeeds only in limiting the meaning of Islam to a narrow, partisan definition: the party itself becomes the location for "true" Islam, and everyone outside it is either totally or partly in the wrong. From this approach comes what ought to be described as "exclusivist" Islam, where different groups claim to be the real spokespersons of Islam and accuse others of being "enemies" of the faith or, more simply, "bad Muslims."

Writing in the movement's magazine *Al-Da'wa* in 1994, Sayyid Nūḥ, another prominent personality in the Muslim Brotherhood, identified ten factors which make it impossible for any world force to stop the growth of his organisation, supporting his claim by citing a number of Qur'anic verses and sayings of the Prophet. His aim in so doing was to show that the Muslim Brotherhood were the best witnesses for Islam, and to assert that because of this "*their* God will not let them down for one moment"[3] (my emphasis).

For other Islamists who defected from the *Ikhwānī* trend and formed the Party of Liberation [*Ḥizb al-Taḥrīr*] in the early fifties in Jordan, "the truth" was viewed somewhat differently. In their opinion, all Islamic movements that had preceded them had failed, "and it was natural that they had failed for, despite being Islamic, these movements were unable to understand the essence of Islam, thus making the situation worse and the problem more complicated. In fact, they were driving society away from Islam rather than implementing it."[4] The point here is not about the validity of the claims of any of these groups, but about the exclusivist perception that each of them maintains: that *they* alone are real Muslims and the real holders of religious truth.

The nature of this partisan religious approach leads its advocates to fight on two simultaneous fronts: on the one hand they claim the only legitimate understanding and representation of Islam for themselves, while on the other they deny this to others, directly or indirectly, so that their mission alone can be justified and glorified. This partisan approach reaches its peak within the *taẓīm*, or organisation, itself, where the organisation becomes an alternative to both society and the state.

From its early days, the Tunisian Islamist movement also developed in line with this trend of thought, therefore it was not surprising that a large proportion of its literature focused on its divine and missionary responsibility in a society that had somehow departed from Islam. The movement's understanding of "true Islam" is marked by symbols and references that are categorically different from those of the Tunisian state educational system, for example. Ghannouchi protested in as early as 1973 about the form of Islamic philosophy that was being taught in Tunisian secondary schools at that time:

> What idea will our pupils get of the teachings of Islamic thought when they cannot find—in the whole text book—even a single quotation from one of the con-

> temporary Muslim thinkers such as Muḥammad Iqbāl, Abū al-A'la al-Mawdūdī, Abū al-Ḥasan al-Nadwī, Ḥasan al-Bannā, Sayyid Quṭb, Muḥammad Quṭb and Malik Bennabi? ... Will the pupils not think that Islam has ceased to exist, that it is no more than a part of our heritage, and has nothing to do with solving our social, economic and political problems? [5]

This "non-existence" of Islam is seen not only as an educational problem for the Islamists; indeed it is much wider. Even in 1981, when they arrived on the political scene, Ghannouchi went to the point of denying the existence of an Islamic society in his country:

> The reason it is necessary to advocate Islam in our society is because of the striking contrast between the brilliant model of Islam and the very low, dark, material reality; the immorality, the cultural Westernization, the social and political injustices, and the alliance—not with Allah, the Prophet and the Muslims—but with the international forces of evil. In conclusion, the Muslim is a stranger in this society; and he feels more of a stranger as he increases his Islamic knowledge.[6]

On forming their political party, their message of the Tunisian Islamists to the public and the political class was that they wanted to play by democratic rules, something that Ḥasan al-Bannā had earlier rejected in Egypt. Was this, then, a sign of moderation or modernism on the part of the Tunisian Islamic movement? It may indeed have been a sign of both, but the exclusivist approach was again too apparent to be hidden by such new terminology. Ghannouchi and his colleagues implied at the party's first press conference that the existence of an Islamic society was not yet a reality, but was rather a goal to be aimed at. Ghannouchi's rationale was as follows:

> We don't want to get rid of one dictatorial tutelage (*wiṣāya*) [simply] to bring another one in its place. Our way is to present ourselves to the people, and if they then choose the Islamic programme, what power could deny them their choice, other than by cancelling their freedom? Similarly, if they choose the communist, or any other ideology, who could oppose the people's will other than by dictatorship?[7]

This is, ostensibly, a very "democratic" message, with the exception of one important point: it links the so-called "Islamic" choice, in a society which to all intents and purposes had been "Islamic" for many centuries, to one small Islamic movement that was at which time only 10 or 11 years old. Here also is Ghannouchi's rather broad definition of the stage at which an Islamic society will be founded: "If the Islamic enlightenment is fruitful and the masses accept Islam to rule their affairs, then Islam will gain its state, and it will be the duty of this state to obey the rules of Allah, and assume its responsibility to implement justice and forbid unfairness."[8]

It may be argued that it was Bourguiba's far-reaching secularist policies that had indirectly strengthened the Islamists' partisan feelings, and the perception that only they could preserve their country's Islamic identity. Here follows a less ambiguous statement by Ghannouchi in an interview given in 1981 to the Kuwaiti Islamists' official magazine: "Until now, the Islamic Trend of Tunisia has not demanded an Islamic state. We do not see this as the movement's goal at the moment. The implementation of the Islamic state must come through the Islamic Trend. It would be a mistake to ask the other parties to fulfil our aims and then to implement Islam."[9]

Tunisia was, of course, already a "Muslim" state in that Islam had been the dominant element in the political and legal system for centuries. The Tunisian people or their rulers' commitment to the directions of Islam may have weakened from time to time and for various reasons, but never to the point that Tunisia could no longer be considered an Islamic country, even with the declaration of the Constitution of the Republic in 1957.

What was it, then, that led a graduate of philosophy and a group of his friends to believe that only they could bring about a truly Islamic state in Tunisia? Again, it reflects the manner in which Islamists assess contemporary Muslims, and the theory of the new, "sectarian" approach to Islam. The Islamists saw themselves as the only party able to defend the Islamic state because, as in the time of the Prophet Muḥammad, they considered the society around them as living in *jāhiliyya*—the term by which the Qur'an described pre-Islamic Arab society—thereby suggesting a state of ignorance and *kufr*, the very opposite of Islam.

The Meaning and Implications of *Jāhiliyya*

When Ghannouchi wrote in *al-Ma'rifa* magazine in July 1979, arguing for theological education as a vital priority for the Islamic movement, his reason for this was that "the individual Muslim living in these *jāhilī* societies is under various pressures," giving as his explanation of "*jāhilī* societies": "Those ... which are not founded on the basis and values of Islam and which do not comply with its laws and teachings."[10]

However, neither the reintroduction of the concept of *jāhiliyya* nor its definition within the contemporary context was an innovation by Ghannouchi. The term was first used by Mawdūdī in respect of Muslims in Pakistan, and was reinforced in the Arab world by the famous martyr of contemporary Islamists, the Egyptian Sayyid Quṭb. Quṭb began his public career as a writer and critic of literature, and joined the Muslim Brotherhood movement in the late forties, becoming implicated in their disputes with President Nasser. He was twice jailed for long terms, and was finally executed in 1966 after being accused of plotting to overthrow the régime. He wrote some of his most influential books in prison, including many chapters of his interpretation of the Qur'an, *Fī Ẓilāl al-Qur'ān.*

A gifted writer with an eloquent style, Quṭb's ideas were further enhanced and glorified after his martyrdom, to the point that he became much more influential after his death than during his lifetime.

In the course of his struggle against Nasser, the champion of Arab nationalism, Quṭb concluded that the circumstances in which Islam had emerged were being repeated in twentieth-century Egypt, reviving and reinterpreting the ancient term *jāhiliyya*. He gave reference to verses 48–50 of *sūra* 5, which conclude: "Do they seek after a judgment of the age of ignorance (*jāhiliyya*)? But who, for a people whose faith is assured, can give better judgment than God?," and argued the following:

> *Al-jāhiliyya* is not a part of history, it is rather a situation which can be found whenever its conditions are present in any place or system. It is, in essence, one which gives power and legislative effect to the wishes of human beings and not to the system of Allah and His laws [by which] to rule life. It does not matter whether these wishes are of an individual, or a class, or a nation, or a whole generation of people; as long as they do not obey the *sharī'a* of Allah, they are all merely subjective desires.[11]

Quṭb then explained further, commenting on verse 50:

> The Qur'anic text guides people to the crossroads: it is either the rule of Allah or the rule of *al-jāhiliyya*, nothing in between and no other alternative is available.... *Al-jāhiliyya* is the state where people are ruled by the laws of people, not of Allah. It has happened before, it may happen today and tomorrow, and it is always contradictory to Islam and in opposition to it.
>
> As for people, any time and anywhere, they either live by the *sharī'a* of Allah without ignoring any part of it, so they can belong to Allah's religion, or they live by a *sharī'a* made by a human being ... and belong to the religion of whoever's *sharī'a* they obey. In no way can they belong to Allah's religion.[12]

Quṭb made it clear to every Muslim that he or she had to choose between "Islam or *jāhiliyya*, *īmān* or *kufr*, the rule of Allah or that of *jāhiliyya*," adding that without taking a decisive position on this issue, a Muslim "will never advance one single step in the right direction."[13]

If the wording of these quotations is seen to echo the similar stand taken by Mawdūdī regarding secularism, then the resemblance is in fact wider than one of mere vocabulary. Although he was the most eminent Arab writer to present a complete theory about the modern *jāhiliyya*, in many ways Quṭb reproduced the ideas of both Mawdūdī and his devoted disciple, Abū al-Ḥasan 'Ali al-Nadwī.

It was Mawdūdī in fact who introduced the use of the term *jāhiliyya* in the context of contemporary Islamic societies, as is detailed in many of his books and lectures. However, it was the translation into Arabic of al-Nadwī's book

Mādha khasira al-ʿālam bi inḥiṭāṭ al-muslimīn? [What did the world lose by the decline of Muslims?] that exported the idea of a "modern *jāhiliyya*" to Arab writers. Quṭb himself wrote a preface to the second edition of al-Nadwī's book in 1951, in which he noted the use of the term *jāhiliyya* to describe the state of all humanity after it had lost an Islamic leadership, and praised the term as being very "precise," reflecting the "author's understanding of the difference between the spirit of Islam and the spirit of the materialistic trend which dominated the world [both] before Islam and after its contemporary decline."[14]

Adopted by Mawdūdī and reinforced by Quṭb, the concept of *jāhiliyya* has been used by many other Islamists all over the Islamic world as a term and method by which to describe the state of contemporary Muslim societies and to decide how to go about reinstating the ideal Islamic model.

It is true that among Tunisian Islamists the term was not widely used, and that even Ghannouchi employed the term only on rare occasions, although the notion of *jāhiliyya* was in evidence from the movement's ideological beginnings. Regarding themselves as modern messengers for Islam in a misdirected or *jāhilī* society, they were strongly motivated to spread their message and expand their ideas and influence. When later subjected to government persecution, they also gave reference to a *jāhilī* society, that is, doing to Islamists what it did before to the Prophet of Islam. In doing so, the fortunes of Islam in fact became identified with those of the Islamists, for they became the only group that was genuinely prepared to fight the "darkness" of modern *jāhiliyya*, both regionally and world-wide. As Ghannouchi wrote in 1979:

> During the fourteenth century AH, Muslims have faced many bitter defeats on all levels. But, thanks to God, they have not given up, and have remained defiant, refusing submission and making sacrifices under the leadership of Islamic movements, until Allah rewarded them with strong signs of victory [the Iranian Revolution]. What we believe is that Islam will continue its costly jihād into the fifteenth century, defending its values and stances. Furthermore it will assume a position of attack in order to liberate the world from the darkness of materialism and its injustices and exploitation, and to build a world in which there will be no conflict between the human being and his conscience. He will be reconciled with religion, as will economics and politics with moral values, today's life with the day after, the human being with his brother human being, all under the *sharīʿa* of Islam, the *sharīʿa* of security, justice and freedom.[15]

The point that this quote illustrates, however, is that an exclusivist approach to Islam in fact leads Ghannouchi to rewrite history and misrepresent the present. It is well known, for example, that those Islamic movements praised by Ghannouchi (the Muslim Brotherhood and *al-Jamāʿa al-Islāmiyya* in Pakistan) played only a minor role in the struggles for national liberation in most of the Islamic world. Whether in North Africa or in the Middle East, national movements were led by the likes of the Front de la Liberation Nationale (FLN) in

Algeria, the Neo-Destour in Tunisia and the Wafd in Egypt. However, his somewhat cavalier approach to historical facts shows how he has given priority to raising morale among his supporters and to portraying Islamic movements as the champions of liberation and progress.

In a society seen as being sick and misguided, the Islamists chose to invest in their *taẓīm*, or organisation, which was seen to be the only alternative in the absence of a truly Islamic society, and a parallel to that which they refused to consider Islamic. The concept of *jāhiliyya* is also one of defensiveness, meaning that the individual Muslims feel alienated, despite being in their own country and among their family and friends, as the leader of *al-Nahḍa* warns:

> [The Muslim] is like a fish out of water.... He faces what amounts to brainwashing because of the domination of the *jāhilī* media. Using their vast powers over the masses, the media try by different means to make him unsure of himself, of his [Muslim] brothers, his religion, the active Islamic groups and their history, and the pious and militant scholars of Islam. At the same time, the media try to remove any psychological obstacles erected by Islam and its culture between Muslims and the rest of *jāhilī* cultures, groups, values and ideologies.[16]

The terms "Islam" or "Muslim" as employed in this quote are used according to the Islamists' exclusivist definition; what concerned Ghannouchi was that his followers would succumb to the misleading picture of Islamic movements and their leaders—in his opinion the *true* representatives of Islam—as projected by hostile media. In this context his young supporters are advised to cut themselves off from society and show total commitment and obedience to the "pious and militant scholars of Islam."

The role of the *taẓīm* was to offer protection from such *jāhilī* pressures, and accordingly, every ordinary Muslim who sought this form of protection became an "Islamist." This differentiation between "Muslim" and "Islamist" is one of the most important pillars of the literature of contemporary Islamic movements, and will be considered in further depth below.

The "Muslim" and the "Islamist"

When Ghannouchi spoke about *jāhilī* societies, he spoke also about its victims: ordinary Muslims. These are not considered *jāhilī* people as such, but their society is *jāhilī* because it is not ruled by the laws and rules of Islam. However, it is not the "Muslim" who will bring about a return to the aims of Islam but the "Islamist." According to Ghannouchi, Islamists are "all those who want to build the human being, civilisation and society on the basis of Islamic values, as they understand them."[17] They are also the real representatives of Islam itself, as Ghannouchi claims: "We are the leading élite of the one million Muslims yearning for progress, civilisation and world peace under Islam. If we are to be mocked and insulted, our enemies should be clear and frank enough to mock and insult Islam itself. Only Islamists provide the real vision."[18]

Such is the essence of exclusivism, upon which the current wave of Islamism has insisted. On many occasions they have denied that they regard themselves as the official spokespeople of Islam, but on many other occasions they have said and practised the opposite. Ghannouchi is clear in his argument that to mock Islamists means to mock Islam, and to attack them means attacking Islam, for they and Islam are one.

It is true that every political group tends to present itself as the best to serve the nation, and it may claim that it is the only real democratic, or socialist or liberal group in the country; however, this differs from a religious claim. There are no serious implications in the traditional rhetoric between competing political groups, but religious claims may lead to dangerous consequences in that Muslims may consider their opponents as enemies of Islam itself, and not merely opponents of a certain political party.

Only Islamists are eligible for membership in this new, self-proclaimed party of "first-class" Islam. Others, referring to the millions of nominal "Muslims" are the victims of the new *jāhiliyya*; they may not be considered enemies of Islam, but are at best ignorant followers of an inferior, "second-class" form of Islam.

The logic behind this new division between "first-" and "second-class" Islam is explained by Ghannouchi:

> The ordinary Muslim may not know his religion well enough to understand the political and civilisational implications of Islam, and this is why we need the Islamists to work for the whole Islamic entity. An ordinary Muslim should theoretically be an Islamist, meaning that his analysis of politics, society and civilisation must be based upon Islam, but this level is not easily reached by people because of the various difficulties in human life. We end up then with a gap between the ordinary Islamic "minimum," and the ideal Islamist model.[19]

For many Muslims, however, such sectarianism within the Islamic community is simply unacceptable, and most, if not all, opponents of the Islamists refuse to accept that a 20-year-old political party can be the sole representative of their faith, or that it is even sincere in its religious claims. Indeed, many people regard the Islamists' position as merely a political manœuvre masked by religious slogans, in an attempt to gain wider popular support.

In the heat of their arguments with secularists and others, Islamists in various countries have tried to defend their adopted title, arguing that "Islamist" does not necessarily exclude others from Islam itself. Indeed Ghannouchi has compared it to the nationalist, socialist and liberal parties, and has argued that one may still find nationalists outside the nationalist party, and socialists outside the socialist party.[20] However, the case with the Islamists is different: "Islamist" is not merely an arbitrary choice of name, but the result of a well-defined ideology which condemns Muslim society for abandoning Islam. Secondly, being religious in essence, it cannot be compared to a political trend such as that of the liberals, nationalists or socialists, for it is not a sin to reject any of these contemporary trends, whereas to be seen to reject Islam is potentially punishable from a religious perspective.

Thus ordinary Muslims in the above position may find themselves the victims of the "sincerity and courage" of someone who sees himself as a "good Muslim," and who wants to go to paradise by killing an "enemy" of Islam, as is currently happening in Algeria and Afghanistan. In this respect, a change of name would not make a great deal of difference when the logic behind it does not change. This is why one continues to read statements by Ghannouchi praising his movement as the only real representative of Islam and condemning his enemies as the foremost threat against it. Even on the occasions when some of his colleagues have criticised him for what they have considered to be political miscalculations, he has used the same line of thought in response, advising them "not to doubt the support of Allah, and to regain faith in their movement and leadership."[21]

It is predictable, therefore, that within this context a majority of those interested in the issue of Islamism, whether researchers or politicians, have vehemently opposed this twofold division within Islam. "Islamism in the dictionary derives from Islam," argues the French orientalist Maxime Rodinson, adding "the reader may risk confusing an excited extremist who wants to kill the whole world, with a normal person who believes in goodness according to the well-respected Muslim concept."[22]

Islamism or Fundamentalism?

The question of "appellation" is by no means simple. It is a part of the political and theoretical debates which have arisen with the resurgence of Islamic movements in both the Arab and Western worlds. It is worth mentioning here that the term "Islamist" may be traced back as far as to the tenth century AD, when the famous Sunnī theologian, 'Alī ibn Ismā'īl al-Ash'arī, wrote this short preface to his work *Maqālāt al-Islāmiyyīn wa Ikhtilāf al-Muṣallīn* [The Sayings of Islamists and the Differences between Worshippers]: "After the death of the Prophet, peace be upon him, people disagreed about many issues; they became different groups and divided parties, but Islam still united them and covered them all."[23]

There was evidently no argument then about the term; it was seen to derive from the word Islam, and had none of the implications accorded to it today. The Islamic community in al-Ash'arī's time was still united, strong and able to absorb conflicting views on policies and doctrines. "Between Sunnīs and Shī'as, and between various divisions of each," argues Hourani, "... was a sense of community, based on the profound conviction of Muslims that to live together in unity was more important than to carry doctrinal disputes to their logical conclusion."[24] This sense of community also extended to Christian and Jewish peoples, "who were recognised as 'People of the Book', who believed in God, the Prophets, and the Day of Judgment, who possessed an authentic revelation and so belonged to the same spiritual family as the Muslims."[25]

In contrast, there is a sharp difference in the current use of the term "Islamist." The prevailing situation in the Islamic world is quite the opposite to that of al-Ash'arī's time, being one of division rather than unity. Islamists now strongly believe that they are different from "ordinary Muslims," and see themselves as an élite that should be entrusted to look after both spiritual and political life. Other Muslims—most importantly those in power—do not agree, and accuse the Islamists of using Islam for political ends. Curiously, however, as governments have begun to realise that religious affiliation is an effective route by which to gain political power, they have felt obliged to use the same logic in an attempt to defend their interests. Accordingly, *they* have begun to portray themselves as the real protectors of Islam, and to portray Islamists as "terrorists" and "enemies of Islam."

Thus *al-Nahḍa* found that its fiercest opponents were not in the West but at home, within those intellectual schools and political parties that insisted that any political group bearing the Islamist tag should be banned, arguing that the whole nation is Muslim and that this is not the privilege of one single party. This is how one Tunisian writer and artist, Izz al-Din al-Madani, has argued against them: "Recognition of these groups who shout that they are "Islamist" would mean that the rest of the society is not Islamic, that is to say *kāfir*.... How is it possible to recognise them and their missionary function for Islam in an Islamic society? Is this not one of the paradoxes of these times?"[26]

The argument was repeated everywhere: "When you use the name 'Islamist'," said one secularist Muslim journalist in an interview with Ghannouchi, "you create a lot of confusion on the political scene and among the general public, because you imply that those who are not members of these [Islamic] movements are not actually Muslims."[27]

The Tunisian Islamist movement found itself with a theological and political problem, and with no past precedent to guide it. Before independence from French rule in 1956, there had been no such division on the Tunisian political scene between Islamists and non-Islamists: all groups, with the exception of the somewhat ambiguous communists, recognised Islam as the most important factor in the country's cultural, religious, and political identity. Furthermore, their attachment to Islam had been a prime dimension of the struggle for national liberation; it was seen by all political parties (again, with the exception of the communists) as both a conviction and a political strategy, and the sole means of safeguarding Tunisians from assimilation and losing their identity.[28] There were, of course, different trends and views among those parties concerning many issues, but they were not articulated as conflicts between "Islamists" and "non-Islamists." The most common terms used to express the different views on the role of Islam in society, especially after independence, were "modernism" and "traditionalism."

Even ex-President Bourguiba, a secular and modernising leader, was keen to present his ideas as being compatible with Islam in so far as it was possible, and was cautious to avoid forcing secularism on all aspects of political and social life. Indeed Hisham Djait, one of the leading contemporary historians on Tunisia, argues that Bourguiba's secularist views were not of a category that may be described as "anti-Islamic."[29] In line with this way of thinking, both Bourguiba and his successor, Zein al-Abidine ben Ali, explained their refusal to legalise any religious political party on the grounds that all Tunisians were Muslims, and that no political party should use a common religious ground to achieve its partisan political aims. Islamists have always rejected this argument, and have accused the government of violating their political and democratic rights.

"It is necessary then," warns the French writer Michel Camau, "not to confuse a politico-religious phenomenon of a circumstantial character with a culture of more than a thousand years;" asserting further, "Tunisian society is Islamic. This is a fact that the events of a decade cannot challenge or deny."[30]

As a consequence of such arguments, Islamists have been widely denied the recognition they so desire. Their movements have been banned in many parts of the Arab world, and even when allowed to operate legally, their "Islamist" appellation has always been in question. Their opponents prefer to call them "*al-uṣūliyyūn*," or "fundamentalists" in English, "*les intégristes*" in French. In the eyes of their enemies they are not even that, but are "extremists" and "reactionaries."[31]

Most Arab intellectuals admit that "fundamentalism" as well as "*intégrisme*" are both Western terms that were originally used in specifically Western circumstances. According to the Oxford dictionary definition, fundamentalism is the "strict maintenance of traditional Protestant beliefs," while the Longman dictionary defines it as "a belief in the literal truth of the Bible," or "a Protestant movement in the twentieth century emphasising such belief."

The term "fundamentalism" does not appear in the *Encyclopaedia of Islam*; the derivative term "*uṣūl*" speaks of roots, or principles. Among the various terminological uses of this word, two are prominent as labels for branches of Muslim learning: *uṣūl al-dīn*, and *uṣūl al-fiqh*. With regard to the French term "*intégrisme*," it is described in the 1975 edition of the *Lexis dictionnaire de la langue française* as: "a tendency of Catholics who claim their adherence to the sole orthodox (or traditional) doctrine, and who wish to maintain the integrity of their doctrines," or an "attitude of those who refuse to adapt their doctrines to new circumstances." Similarly, the 1932 edition of *Le Petit Robert* says that "*L'intégrisme* is a doctrine which tends to maintain the totality of a system, or an attitude of Catholics who refuse all kinds of evolution"; the "*intégriste*" is described also as a member of a Spanish party which called for the submission of the state to the church.

Thus we may see that both the English and French terms used to define modern Islamic movements are by no means neutral terms, rather, they have negative connotations rooted in Western history, in that they are linked to certain orthodox religious trends that have refused evolution and change. When Western analysts began to pay close attention to Islamic movements following the Iranian Revolution in 1979, they found terms such as "fundamentalism" and "*intégrisme*" to be the closest descriptions that could be applied to what they knew of autocratic religious trends in their own history. Thus they adopted the two terms and made them universal in their application.

For the United States in particular, which had backed the Shah's falling régime in Iran, almost everything connected to Islam at that time was seen in a negative light, and not only that relating to Islamic movements and their activities. Coverage of Islam from the Western perspective—what Edward Said has coined the "covering up" of Islam—was seriously imbalanced:

> This coverage—and with it the work of academic experts on Islam, geopolitical strategists who speak of "the crescent of crisis," cultural thinkers who deplore the "decline of the West"—is misleadingly full. It has given consumers of news the sense that they have understood Islam without at the same time intimating to them that a great deal in this energetic coverage is based on far from objective material. In many instances "Islam" has licensed not only patent inaccuracy but also expressions of unrestrained ethnocentrism, cultural and even racial hatred, deep yet paradoxically free-floating hostility. All this has taken place as part of what is presumed to be fair, balanced, responsible coverage of Islam.[32]

Below is another definition of "fundamentalism," as seen from a wider, universal perspective:

> Religious fundamentalism, whether Jewish, Christian, or Islamic, consists of an assertion that the received sacred text consists of a set of eternal living truths. It has, therefore, an in-built guarantee of infallibility. It is an ideology of the Books as the all-sufficient guide in every condition and circumstance of life in whatever century or for whatever purpose. It neglects transcendence and open-endedness and avoids the need for a creative re-interpretation of the faith. It sees God in the light of its own concepts. Having made up its mind what the faith should be, isolated texts are then used as proof texts, often in support of some particular cause.[33]

This definition does not totally explain the notion of fundamentalism, however, in that it is also an ambitious theory, an activist trend which expresses a religious obligation to save the sacred (and unchanged) faith and doctrines from any distortion. The implications of this, as presented below, appear to be somewhat ominous:

> The notion of activism is, therefore, closely associated with that of fundamentalism; and some movements may behave in a manner which justifies the use of the term "militant," or even "terrorists." But these are far from being essential characteristics. Similarly, the assertion that one set of beliefs is true can lead to a denial of the veracity of all others, and then intolerance and persecution may result. Fundamentalism is clearly more likely to produce an atmosphere of confrontation rather than co-operation, but it need not inevitably do so.[34]

Twice in this passage the author attempts to dilute the dangerous liaisons he suggests between fundamentalism on the one hand, and terrorism and intolerance on the other, arguing that these characteristics are "far from being essential" to fundamentalism. However, this is little comfort to those regarded as fundamentalists, who have already been labelled with a series of negative descriptions: mainly "conservative" and "anti-evolutionist," besides which they are usually portrayed as dissidents and terrorists. Even Roger Garaudy, the French philosopher who converted to Islam, affirms in the first paragraph of his book *Intégrismes* that "*intégristes*, all *intégristes*, whether technocrats, Stalinists, Christian, Jewish or Islamic, today represent the biggest danger for the future."[35]

A minority of Western thinkers disapprove of this general trend of condemning Islamic fundamentalism as one entity; two well-known proponents of this trend are the French scholar François Burgat and the American John Esposito. Both have met Ghannouchi many times and have written sympathetically about both him and his movement. In his book *The Islamic Threat*, Esposito has criticised orientalists such as Bernard Lewis and all those who see "the Muslim world through the prism of Khomeini and revolutionary Iran," arguing that their work has deeply influenced American perceptions of Islam and the Middle East, and that it has "often obscured the differences and divisions in the Muslim world; the many countries and Muslims that did not follow his lead; Iran's failure or at best very limited success in exporting and inciting Islamic revolutions; and the many voices other than Khomeini's who spoke out on Salman Rushdie."[36]

The general direction of this debate about Islamism and fundamentalism is not in favour of the Islamists. For many of them, the consensus among many Western and Arab writers about the "dangers" of fundamentalism is another sign of the "unholy" alliance between the West and Arab secular élites. Islamists themselves disapprove of the use of the translation of the Arabic word *al-uṣūliyya* as "fundamentalism," and similarly reject its attendant implications. However, they have not been able to counterbalance effectively either the influence of the mass media or the insistence of Western and Muslim secular writers on imposing what the Islamists see as an inappropriate and misconceived concept of them. This is why some Islamists have opted for a more pragmatic stance, and although accepting the term, have tried to change its negative connotations. 'Abd al-Majid al-Najjār of *al-Nahḍa* explains: "*Al-uṣūliyya* derives from *uṣūl*, meaning the Islamic trend that accepts the basics of Islam [the Qur'an, traditions, views of the *Ṣaḥāba* as the basis of their ideological, political and social programmes."[37] He also extends the scope of the term to cover all those who believe in the Islamic identity of Tunisia, as accepted in its long history since the country was Islamised at the end of the seventh century CE.[38]

As for Ghannouchi and the movement itself, they found that problems regarding the movement's name had done them more harm than good as far as their political interests were concerned. Thus in 1988 the movement dropped its historical name, *Ḥarakat al-Ittijāh al-Islāmī*, and chose its current name, *Ḥarakat al-Nahḍa*, in which no "Islamic" term is included. Although this was largely a political move taken to obey the regulations banning "Islamic" political parties, the movement's theoreticians also attempted to rationalise it. Ghannouchi explains that although "Islamists" is in fact the best possible term for those seeking the Islamic model, such a term is not essential, because names do not necessarily convey the truth. "We gave up the term on that basis," he says. "If we had believed that it was essential, we would not have dropped it."[39]

It is ideas, not names, that concern the Tunisian Islamists: even if they have given up the term "Islamic" in their official name, their ideology, projects and aims remain unchanged. They still maintain that their society is "sick" and in need of Islam—that is, an Islam in its comprehensive sense—which we shall now examine in more detail.

The Principle of Comprehensiveness

One central point in the ideology of contemporary Islamic movements, *al-Nahḍa* included, is the firm belief in the validity of Islam, not only to guide society in matters of faith and culture, but also in all political, social and economic aspects of life. All movements begin by asserting that Islam is both a religion and a state (*dīn wa dawla*), meaning that it must rule in matters of faith as well as political issues. They all believe that Islam is "more than just a matter of faith and ritual and that it provides solutions to all social, political, economic and cultural problems," and argue that "the present economic, political, military, social and cultural weakness of the Muslim world is the direct result of deviation from the true path, and in particular the espousal of secular and materialistic ideologies and values."[40]

The two main notions that Islam is comprehensive and that it is the only solution for the current weakness of the Islamic world, provide all Islamists with an ideology, an aim and a *raison d'être*. When they proceed to the details of this basic theory, however, they differ among themselves on how to bring about the true Islamic state, and thus become divided into different groups and parties. As has been mentioned, *al-Nahḍa* belongs to the mainstream Muslim Brotherhood trend, which supports a view of Islam as defined by Ḥasan al-Bannā:

> We believe that Islam's regulations and teachings are comprehensive. They organise people's affairs in this life and in the next. Those who believe that these teachings concentrate on theological and spiritual matters are simply mistaken. Islam is belief and worship, country and nationality, religion and state, spirit and action, Qur'an and a sword. The Holy Qur'an is very clear about all this and considers it part of the essence of Islam.... The Muslim Brotherhood is totally convinced that Islam has this overall comprehensive meaning, and that it must rule and supervise all aspects of life ... as long as the *umma* wants to be Islamic in the correct manner.[41]

Because of the early influence of the Syrian Muslim Brotherhood on Ghannouchi, his understanding of Islam and the duties of the Tunisian Islamists is close to that of his Syrian counterparts. From the days of the organisation's beginnings in 1969 as *al-Jamā'a al-Islāmiyya*, the Islamists promoted themselves as representatives of the Muslim Brotherhood's ideas in their country. In turn they became part of the international network of Muslim Brothers, and every full member was asked to swear to serve the movement as part of the international organisation of *al-Ikhwān al-Muslimūn*. This rule was still valid in the mid-1990s.

With the formation of the MTI in 1981, its leaders announced their manifesto, in which they stated that there were two main factors behind the emergence of their trend—one religious, the other political. On the religious side, they stated that Islam had become little more than a symbol encircled by cultural, moral and political dangers, being the direct result of a prevailing lack of interest and attacks on its values, institutions and people.[42] On the political side, it was argued:

> Despite gaining its formal documents of independence, Tunisia suffered from special circumstances characterised by crisis, increasing social confrontation and the suspension of efforts for overall development. This situation was further exacerbated because of the unitary political control of the Destour party, and its growing inclination towards total domination of power and mass institutions and organisations. It was also due to impromptu political and social policies and its fluctuations and dependency on international interests, which contradict the national interests of our people.[43]

Because of these two main reasons, the Islamists felt obliged by their "divine, national and human responsibility, to continue their efforts and improve them, aiming for the real liberation of the country and its progress on the just bases of Islam and its straight path."[44]

The combination of the religious and the political was strongly emphasised in every part of the 1981 declaration, and was clearly marked as representing the basis of the movement's ideology and its Islamic identity. The Islamists prepared themselves for a negative reaction, as this defiant excerpt shows:

> Some will describe our activities as a way of mixing religion with politics. This criticism not only reflects a clerical approach alien to our original culture, but also suggests a desire to prolong the historical deviation of the Islamic world.... The [members of] the Islamic Trend Movement, without claiming to be the official spokesmen for Islam in Tunisia, think it is the Movement's right to adopt a broad approach to Islam, which forms the basic theory from which come the various intellectual views of the movement, as well as its political, economic and social programmes, which shape the identity of the movement and establish its strategic and tactical positions.[45]

To match this analysis, the first goal of the MTI—out of the five mentioned within the declaration—was to bring about the rebirth of Tunisia's Islamic identity, so that it could both reclaim its status as a great base for Islamic civilisation in Africa and bring an end to its "crisis of identity." The second goal consisted of the renewal of Islamic thought on the basis of the fundamental principles of Islam, in addition to purging it of the traces of the age of "backwardness" and the influence of Westernisation. In the list of means by which to achieve those goals, the use of mosques to mobilise the masses assumed first place.[46]

In many ways, the declaration reflects closely the intention of *al-Jamā'a al-Islāmiyya* some ten years earlier to serve Islam and build a true and just Islamic state, in opposition to the modern, secular Bourguiban version. Then, as now, the Islamists were compelled to present a strong front on three main issues: firstly, to justify the inclusion of religion in politics in accordance with the original texts of Islam, thus making their case against secularism; secondly, to justify their call for an Islamic state in an already-Islamic country that had adopted Islam in its constitution as its official state religion; and thirdly, to present a detailed model for the Islamic government that the movement so hoped to bring about.

Islam and Secularism

In arguing for a broad Islam that ruled over both the individual and collective society, *al-Nahḍa* rejected the secularist concept that separated religion and the state. From the outset, Ghannouchi and his colleagues were unequivocal in their aim of placing religion at the heart of the political system. From as early as 1973, Ghannouchi expressed his confidence in what he saw as the solution for the problems of Tunisia and the whole Muslim world:

> In the past, our *umma* could only go forward with Islam.... The main revolutions in our history have been based on Islamic values and that is the reason behind their success. When we tried in other instances to make radical changes and ignored those values, our programmes and plans ended in total failure.

> When we call for the rebirth of Islamic values in our minds throughout the entire *umma*, it is not simply to achieve the *baraka* of religion, or because it would solve our problems on the Day of Judgment before God.... The problem of Muslims is that they do not adopt the Islamic solution to their problems, and this is why they are backward.[47]

In 1979, when the movement began to attract negative attention from the authorities, Ghannouchi responded to those who accused the Islamists of using Islam for their political ends by:

> in principle, we find no problem in insisting that politics is a part of Islam's comprehensive programme with which to conduct life. Islam strongly rejects the Western concept of the separation between religion and state, because it regards the state as a servant to religion, responsible for its protection, the execution of its orders, and making its word the most influential in the world.[48]

In this context, Ghannouchi asserted: "the Islamic state is the main goal for the political movement that adopts a religious basis."[49] Referring to evidence with which to support this theory, he alludes to Qur'anic verses from *sūras* 4 and 5, as is done by most Islamic movements. Here is al-Najjar's response to those who questioned the connection between religion and state:

> It is possible to confirm Islam's authority to rule over political and social life with just one verse, such as: "But no, by thy Lord, they can have no [real] faith, until they make thee judge in all disputes between them" (4: 65), and the verse: "If any do fail to judge *[ḥukm]* by [the light of] what God has revealed, they are [no better than] unbelievers" (5: 44).[50]

However, these verses do not contain any clear instructions or indications about a political system that may be called "the Islamic state," and some Muslims, such as the Egyptian Dr Muḥammad Aḥmad Khalafallāh argue that the Arabic word *ḥukm*, as used in the previous verses, relates to the judicial system only, and not to a system of government.[51] Nevertheless, the political interpretation is still defended in two further arguments by the Tunisian Islamists. The first is by al-Najjār:

> Islamic teachings came to enlighten the human being with the truth about his existence, as well as with the basics for social and economic dealings, also rules for the administration of the *umma*'s affairs, meaning its political affairs.... On the basis of this, the first element in the political dimension of [Tunisian] identity according to the *uṣūlī* approach is that the overall rules for the government should be drawn from the revelation. First among the revelation's rules on government is that the nation must organise itself politically and adopt a presidential institution, which is called in Islamic thought the "Caliphate" or "Imamate."[52]

The second argument is presented by Ghannouchi:

> To emphasise the necessity of government in Islam or for Islam does not mean it is a part of it, because there is no direct order in Islam to install it. However, its absence will make void all—or the majority—of the laws of Islam. As association is necessary for people's life and prosperity, and as they definitely need a government to uphold justice among them, and because Islam is a canon of justice, that government will either be responsible for implementing it and thus being Islamic, or for implementing another canon and thus being non-Islamic; call it what you like, but not Islamic.[53]

The Islamists' approach to the issue of the Islamic state is as valid today as it was at the time of the movement's foundation. There is (in theory) no compromise with any form of secularism, as there is only one formula for an Islamic state, despite there being many others for non-Islamic models. Ghannouchi chose merely to follow and repeat what the main leaders of contemporary Islamic trends had said before him, especially those whom he had selected as the most important reformers of this century: Mawdūdī, Ḥasan al-Bannā and Ayatollah Khomeini.[54]

Al-Bannā's views were widely known in this respect: he argued that one of the main goals of his movement was to set up a

> Muslim government which will guide the people to the mosque and lead them according to the teachings of Islam. We do not recognise any government that is not based on Islam, we do not recognise these political parties, nor these traditional forms of rule imposed upon us by the enemies of Islam. We will work to reinstate the Islamic régime in all its aspects, and form an Islamic government based on this régime.[55]

As for Mawdūdī, he asserted that Islam could never truly be fulfilled without being the sole criterion for the state's affairs. His clear advice to Muslims was as follows:

> If you truly and strongly believe in Islam with which came the Qur'an and Muḥammad, peace be upon him, then you have to fight democratic nationalistic secularism, and take the lead to implement the Lord's *khilāfa*, based on the worship of God wherever you are and wherever you settle, especially in the country where you are in control. But if you choose to be responsible for this régime that does not believe in Allah and his messengers, then there is nothing we can do but to cry and feel sorry for your fake Islam, and your false claim of abiding by it.[56]

For Mawdūdī, secularism was the *opposite* of religion, because the relationship between God and the human may take only one of two forms:

> Either Allah is indeed the creator, master and governor of the human being and the universe in which he lives, or he is not so.... If Allah is the creator, master and governor ... then it is not logical that his rules and jurisdiction should be restricted to the personal affairs of the individual and should not cover the relationship between him and others in public life.... If there were no need for Allah and his teachings in our family life, our community affairs, or in the school and faculty, shops and the market, or in parliament and government, or in the government's administration, or in the military camp ... then what would be left of the meaning of worshipping God?[57]

This categorical rejection of secularism is one of the main themes of Mawdūdī's thought. He shuns secularism because of its capacity to "conceivably exclude all morality, ethics or human decency from the controlling mechanisms of society," and because "morality of any kind is simply inconceivable without religion and the sanctions of eternal punishment to support it," arguing, "when religion is relegated to the personal realm, men inevitably give way to their bestial impulses and perpetrate evil upon one another. In fact, it is precisely because they wish to escape the restraints of morality and divine guidance that men espouse secularism."[58]

The argument concerning secularism in the Islamic world is not, of course, decided by the assertions of the Islamists alone. Many other Muslims, especially those among the ruling classes and the intelligentsia, defend secularism, not necessarily as an alternative to Islam, but as an "Islamic" idea, or at least as a concept that is compatible with Islam. The Algerian Mohamed Arkoun may be seen to represent this trend when he argues for what amounts to a "manifesto" for "Islamic secularism." In his opinion, not only is "secularism included in the Qur'an and the Medinan experience," but also, he asserts, "the Umayyad–'Abbasid state is secularist; ideological theorising by the jurists is a circumstantial product using conventional and credulous arguments to hide a historical and political reality; this theorising is built on an outdated theory of knowledge."[59]

Arkoun also mentions that the "very early military powers played a pre-eminent role in the caliphate, the sultanate and all later forms of Islamic government," and argues, "attempts to rationalise the *de facto* secularism and to develop a lay attitude have been made by the *falāsifa*," which explains why the "new history of Islamic thought has to devote a chapter to the sociology of the failure of philosophy: it is one of the requirements of reasserting a philosophical attitude in Islamic thought."[60]

For the Tunisian Islamists, as well as for the vast majority of Islamists in the Arab world, Mawdūdī's argument is representative of their stand against secularism, while Arkoun's view is dismissed as an apologia for a Western argument, which they oppose. They do not hide their rejection of the Western separation between religion and state, which, in their opinion, clearly contradicts the principles of Islamic political theory. Ghannouchi argues:

> There is no future for a political group that wants to govern on the basis of refusing the rule of Islam. It has no hope of ruling a society with which it cannot be associated, simply because this society is of an Islamic nature ... there are no grounds for our secularists to believe that Islam may abandon its authority to rule and guide society at any time in the future. They have to give up their hopes of changing the nature of Islam to "Christianity," meaning a set of beliefs that bears no relationship to the running of society's affairs.[61]

On this matter Ghannouchi takes an uncompromising stand, and the Tunisian Islamists offer a number of arguments in his defence. Al-Najjār, the second leading theoretician of the movement, argues that the emergence of secularism in Europe was the result of various factors such as "the nature of Christianity, which does not claim to cover all aspects of life," and "the atrocities perpetrated by the Church in the name of religion"; reasons that are deemed void in the Islamic context, since "Islam rules all aspects of human life, and there has been no clergy in its history claiming special religious authority and becoming a dictatorship, ruling the people by force. This is why the Islamic world did not know secularism before its emergence in the West, meaning that it is simply imported from them."[62]

Although there have been a number of Muslim and non-Muslim thinkers who have defended the concept of secularism and have tried to make it compatible with Islam, such as Arkoun and 'Alī 'Abd al-Rāziq of Egypt (1888–1966) before him, in his *Al-Islām wa Uṣūl al-Ḥukm* ["Islam and the Bases of Political Authority"], the majority of writers and experts on Islam do recognise what may be called the "special relationship" between Islam and the state.

It seems in fact that there is recognition of the anti-secularist concept in Islam among even the majority of researchers in the West. The only difference here is that some of them choose to highlight this point as a "negative" feature, especially when addressing a Western public which disapproves of interference of the religious in the political. Bernard Lewis, for instance, claims that there was no distinction between "church" [sic] and "state" in classical Islam, and makes a comparison with Christendom where "the existence of two authorities goes back to the founder, who enjoined his followers to render unto Caesar the things which are Caesar's and to God the things which are God's. Throughout the history of Christendom there have been two powers, God and Caesar, represented in this world by *sacerdotium* and *regnum*, or ... church and state."[63]

Lewis also argues that classical Arabic did not even possess vocabulary corresponding to the notions of "spiritual and temporal ... religious and secular," and that it was not until the nineteenth and twentieth centuries, when the Arab world was under the influence of Western ideas and institutions (in addition to direct colonial rule over many Islamic countries), that new words were found, first in Turkish and then in Arabic.[64]

Other researchers agree on the Western nature of secularism and have argued that it was accepted by some Muslim reformers as a theory for progress:

> In the Islamic world, it was seen by a group of reformers as a tool to face up to the European challenge. This group advocated direct borrowing from the Western models of secular culture and nationalism, the presumption being that the only effective way to meet the challenge of Western political and cultural intrusion was to adopt, at least in part, those institutions that had made the European powers so irresistibly powerful. Also implicit in this approach was the premise that Western secular civilisation was innately superior to the established Islamic culture of the Middle East.[65]

For *al-Nahḍa*, there is no way in which secularism can be compatible with Islam and its culture. It is simply, in Ghannouchi's words, "a foreign product, a Christian product that grew in the period of intellectual colonialism [or conquest] and which has failed to solve any of our nation's problems. People's refusal to accept these ideologies is increasing."[66]

What has made this refusal so strong is a similar radicalism in the position of those Tunisians who advocate the idea of the secular state, most of whom do not argue from an Islamic viewpoint, attempting to show that Islam is "secular" or that it approves of secularism. Rather, many of them (the Marxists in particular), have made their fight for secularism a fight against religion in general.

In a number of his speeches, President Bourguiba proclaimed a new kind of régime and state based on "reason." As we have already seen, when he invited the Tunisian Muslims to abstain from the fast of Ramaḍān in order to maintain their capacity for work and avoid low productivity, he argued that this illustrated how Islam is based on reason.[67] Although very provocative, it is clear that Bourguiba was none the less attempting to find a religious rationale to support his opinion; indeed most of his policies for modern Tunisia were presented and defended as being compatible with Islam and the policies of a Muslim ruler.

Although Bourguiba did not favour the total separation between Islam and the state, he felt obliged to respect deeply rooted Islamic feelings inasmuch as it was necessary to have his ideas implemented. Despite its obviously secularist bent, his official political message was not to break with Islamic ideology completely. As one Tunisian writer argues, the Islamic connection was always required by the Bourguiban régime, "mainly to defend its legitimacy and strengthen it among the traditional classes. Besides, this legitimacy was always needed to confront the Islamic movements, which could challenge the religious values of the government."[68]

As we have seen, Bourguiba was not the only secularist to oppose the modern Tunisian Islamic movement, although as the main holder of power in the country he was its most prominent opponent. A more radical defence of secularism came from the Marxist groups with whom the Islamic movement clashed at the University. For the likes of Hammami, leader of the Tunisian Communist Party, Bourguiba was a servant to traditional Islamic values. He protested at the fact that the Tunisian constitution ruled Islam to be the religion of the state and its president, describing it as "a means to protect the interests of the bourgeoisie and to use religion as an opiate to appease the labour masses."[69]

After arguing that an Islamic state was one of oppression and dictatorship, Hammami suggested that the secular state, for which the Marxist Leninists and all true progressives called, represented the most significant guarantee of freedom of belief in Tunisian society,[70] while Mohamed Maali, who wrote the preface to Hammami's book *Ḍidda al-Ẓalāmiyya* [Against Obscurantism], argued that the truly important issue for contemporary Tunisia was the demand for a secular state: "It is now clear that the demand for secularism or its rejection is the issue that distinguishes the real supporters of democracy from their enemies, even if they mask their intentions with Islamic slogans."[71]

For al-Najjār of *al-Nahḍa*, all calls for secularism, whether oblique as in Bourguiba's case, or official and unequivocal as proclaimed by the leftists, are simply contradictory to the true essence of Islam. His belief is that Islam only should rule over society, while "the aim of secularism," he says, "is to base our policies on education, culture, economy and international relations—not on Islam, but on human reason."[72] He denies that the Islamic state will open the door for dictatorship, for in his opinion this notion comes from a "theological misunderstanding about Islam which has no relationship with the truth of Islam," while arguing that dictatorships under secularist and leftist régimes have been far worse than anything under an Islamic ruler in Islamic history.[73]

Secularism in the Islamic world is of course not merely a simple target either to attack or to eliminate; it was and still is a complicated political, economic, social and cultural process which will prove to be very difficult to reverse. Hourani provides a revealing account as far as this process is concerned:

> In the realm of law, civil and criminal, commercial and constitutional, secular codes have replaced the religious (from the mid-1940s and 1950s onwards) everywhere, except in the most withdrawn parts of Arabia, and this has happened less completely in a country such as Egypt, where it has taken place almost imperceptibly, whereas in Turkey it has been the product of revolution. In Turkey and Tunisia an attack has been made even on the last stronghold of the *sharīʿa*, the law of personal status: polygamy has been abolished and civil marriage introduced, apparently without causing great scandal.[74]

Secularisation is indeed still under way in all parts of the Arab world, and even in Saudi Arabia itself. With a stronger Western influence than ever before, the modern Arab state is more and more inclined to disregard what may be seen as the obligations of Islam, while at the same time overlooking the rudiments of secularism, especially the democratisation of political life and making governments accountable to free elected parliaments. The current trend for ruling régimes is to depend on financial, political and military help from Western powers to remain in control, and to ignore all popular calls for either Islamisation or democracy. As a result, secularism becomes merely a political tool used by many régimes to suppress active Islamic movements, and has nothing to do with democratising political life.

As the Islamists' presence in Tunisia began to grow in the late eighties, their fight against secularism became increasingly difficult on both the intellectual and political fronts. Their political efforts might have been halted as a result of their last confrontation with President Ben Ali, but Ghannouchi and his colleagues continued to present themselves as the leading Tunisian intellectual force for the promotion of democracy and a modern, comprehensive Islam, and formulated what amounted to a detailed theory about the modern Islamic state.

Notes

1. See Ḥasan al-Bannā, *Majmū'at rasā'il al-imām al-shahīd Ḥasan al-Bannā* (Cairo: Kutub al-Da'wa, n.d.), p. 75.
2. Sa'īd Ḥawwā, *al-Madkhal ilā da'wat al-Ikhwān al-muslimīn* (n.p., n.d.), p. 39.
3. Sayyid Nūḥ, in *Al-Da'wa*, 38 (1994).
4. See the Ḥizb al-Tahrīr publication *Principles of Ḥizb al-Tahrīr*, 4th edn (n.p., n.d.), p. 9.
5. Al-Ghannūshī, *Maqālāt*, op. cit. p. 4 .
6. Ibid., p. 174.
7. Extracts from this press conference are located in al-Ḥāmidī's *Ashwāq al-Ḥurriyya*, op. cit., p. 89.
8. Al-Ghannūshī, *Maqālāt*, op. cit., pp. 175–6.
9. Extracts from this interview are documented in al-Ḥāmidī, *Ashwāq al-Ḥurriyya*, op. cit., p. 81.
10. Al-Ghannūshī, *Maqālāt*, op. cit. p. 132.
11. Sayyid Qu b, *Fī ẓilāl al-Qur'ān*, vol. 5 (Beirut: Dār al-Shurūq, 1985), p. 891.
12. Ibid., p. 904.
13. Ibid., p. 905.
14. Quṭb's preface to Abu al-Ḥasan 'Alī Al-Nadwī's *Mādha khasira al-'ālam bi inḥiṭāṭ al-muslimīn?* (Kuwait: IIFSO, 1978), p. 22.
15. Al-Ghannūshī, *Maqālāt*, op. cit., p. 104.
16. Ibid., p .132.
17. Al-Darwīsh, *Yaḥduthu fī Tūnis*, op. cit., p. 16.
18. Al-Ghannūshī, *Maqālāt*, op. cit., p .71.
19. Al-Darwīsh, *Yaḥduthu fī Tūnis*, op. cit., pp. 16–17.
20. Ibid., p. 16
21. Al-Ghannūshī in *al-Murāsala*, an unpublished internal newsletter circulated among members of al-Nahḍa, 1 December 1995.

22. Maxime Rodinson in *Revue Parlementaire*, 928 (March–April 1987).
23. 'Alī ibn Ismā'īl al-Ash'ari, *Maqālāt al-islāmiyyīn* (Cairo: Maktabat al-Nahḍa, 1969), p. 34.
24. Albert Hourani, *Arabic Thought in the Liberal Age* (Cambridge: C.U.P., 1983), p. 4.
25. Ibid., p. 4.
26. Quoted in the preface to Anas al-Shābbī, *al-Taṭarruf al-dīnī fī Tūnis*, (Tunis: La Presse, 1991), p. 5.
27. Al-Darwīsh, *Yaḥduthu fī Tūnis*, op. cit., p. 14.
28. Toumi, *La Tunisie de Bourguiba à Ben Ali*, op. cit., p. 113.
29. As documented in al-Munṣif Wannās's *al-Dawla wa al-mas'ala al-thaqāfiyya fī Tūnis* (Tunis: Dār al-Mīthāq, 1988), p. 132.
30. Michel Camau, "Chronique politique, Tunisie" in *Annuaire de l'Afrique du Nord*, 1979 (Paris: Editions CNRS, 1981).
31. Ḥamma al-Hammāmī, *Ḍidda al-ẓalāmiyya*, op. cit., p. 3.
32. Edward W. Said, *Covering Islam: How the Media and Experts Determine How We See the Rest of the World* (New York: Pantheon Books, 1981), p. ix.
33. A.K. Lambton, *Islamic Fundamentalism*, p. 33.
34. Ibid., p. 5.
35. Roger Garaudy, *Intégrismes* (Paris: Pierre Belfond, 1990), p. 9
36. John L. Esposito, *The Islamic Threat: Myth or Reality?* (New York: OUP, 1992), p. 180.
37. 'Abd al-Majīd al-Najjār, *Ṣirā' al-huwiyya fī Tūnis* (Paris: Dār al-Amān, 1988), p. 31.
38. Ibid., p.31
39. Al-Darwīsh, *Yaḥduthu fī Tūnis*, op. cit., p. 19.
40. Patrick Bannerman, 'Le Mouvement de la Tendance Islamique in Tunisia', op. cit. p. 68.
41. Ḥasan al-Bannā, *Islamunā* [A speech delivered at the *Ikhwān*'s 5th conference] (Cairo: Dar al-I'tiṣām, 1977), pp. 14–15.
42. The founding statement of the MTI is located in Rashid al-Ghannūshī's *Maḥāwir Islāmiyya* (Khartoum: Bayt al-Ma'rifa, 1989), p. 157.
43. Ibid. p. 157.
44. Ibid. p. 158.
45. Ibid. p. 158.
46. Ibid. pp. 158–9.
47. Rāshid al-Ghannūshī, *Min al-fikr al-islāmi fī Tūnis*, vol. 2. (Kuwait: Dār al-Qalam, 1992), pp. 26–7.
48. Ibid. p. 64.
49. Al-Ghannūshī, *Ḥarakat al-ittijāh al-islāmī fī Tūnis*, vol. 3, op. cit. p. 126.
50. Al-Najjār, *Ṣirā' al-huwiyya fī Tūnis*, op. cit., p. 80.
51. Muḥammad Aḥmad Khalafallāh in *al-Ḥarikāt al-islāmiyya al-mu'āṣira fī al-waṭan al-'arabī*, op. cit., p. 90.
52. Al-Najjār, *Ṣirā' al-huwiyya fī Tūnis*, op. cit., p. 39.
53. Rāshid al-Ghannūshī, *al-Ḥurriyyāt al-'āmma fī al-dawla al-islāmiyya* (Beirut: Markaz dirāsāt al-waḥda al-'arabiyya, 1993), pp. 91–2.
54. Al-Ghannūshī, *Maqālāt*, op. cit., p. 87.

55. Ḥasan al-Banna, *Risāla ilā al-shabāb* (Cairo: Al-Zahra for Arab Media, 1989), pp. 60–1.
56. Abu al-A'lā al-Mawdūdī, *Islam and Modern Civilisation* (Cairo: Dār al-Anṣār, 1978), p. 42.
57. Ibid., pp. 19–21.
58. Extracts taken from Tamadonfar Mehran, *The Islamic Polity and Political Leadership: Fundamentalism, Sectarianism and Pragmatism* (Boulder: Westview Press, 1989), p. 38.
59. Mohamed Arkoun, "The Concept of Authority in Islamic Thought," in Islam: State and Society, ed. Klaus Ferdinand and Mehdi Mozaffari (London: Curzon Press Ltd., 1988), p. 71.
60. Ibid.
61. Al-Darwīsh, *Yaḥduthu fī Tūnis*, op. cit., p. 34.
62. Al-Najjār, *Ṣirā' al-huwiyya fī Tūnis*, op. cit., p. 43.
63. Bernard Lewis, *The Political Language of Islam* (Chicago and London: The University of Chicago Press, 1988), p. 3.
64. Ibid.
65. Alan R. Taylor, *The Islamic Question in Middle East Politics* (Boulder: Westview Press, 1988).
66. Al-Darwīsh, *Yaḥduthu fī Tūnis*, op. cit., p. 36.
67. Ali al-Ganari, *Bourguiba le combattant suprême* (Paris: Plon, 1985), pp. 301–7.
68. Wannās, *al-Dawla wa al-mas'ala al-thaqāfiyya fī Tūnis*, op. cit., pp. 114–5.
69. Al-Hammāmī, *Ḍidda al-ẓalāmiyya*, op. cit., pp. 4–5.
70. Ibid., p. 35.
71. Ibid., p. i.
72. Al-Najjār, *Ṣirā" al-huwiyya fī Tūnis*, op. cit., p. 50.
73. Ibid., p. 85.
74. Hourani, *Arabic Thought in the Liberal Age*, op. cit., p. 350.

6

The Islamists' Islamic State

One of the most common accusations made against modern Islamic movements is that they lack specific solutions for the political, social and economic problems of their countries. They tend, as John Esposito claims, "to be more specific about what they are against than what they are for. While all may speak of an Islamic order or state, of implementation of the *sharī'a,* of a society grounded more firmly on Islamic values, the details are often vague."[1]

Since the date of the formation of their political party in 1981, the Tunisian Islamists, when challenged on this point, have always maintained that it was the government's repression that prevented them from working on a detailed political and social programme. Ghannouchi addressed the movement's critics on this point: "You demand from us renewal and want to see our views and programmes and, initially, you have good reason to do so. However, we are occupied with something more important: defending our right to existence and standing up to the efforts being made to suppress our movement."[2]

However, it is remarkable that *al-Nahḍa* in general, and Ghannouchi in particular, have produced an interesting range of literature about the political system they wish to see in place in Tunisia. Since 1981 the Tunisian Islamic Movement has adopted what may be considered a very "democratic" approach to questions of political order and freedoms, thereby gaining respect from many Arab intellectuals outside Tunisia, notably the group of Arab nationalists who published Ghannouchi's book *al-Ḥurriyyāt al'amma fī al-Dawla al-Islāmiyya* [Civil Liberties in the Islamic State]. Ghannouchi has also earned the respect of certain Western scholars such as Burgat and Esposito, the latter of whom has argued

> despite the government's attempt to paint the MTI as retrogressive, fundamentalist fanatics and a violent Iranian-backed revolutionary movement, Ghannouchi … distanced the movement from the excesses of revolution and advocated a Tunisian rather than an Iranian solution. Ghannouchi denounced the use of violence and instead chose to work within the system, emphasising a gradual process of social

transformation and political participation as the means to realise [the] MTI's long-range goal of establishing an Islamic state.[3]

The Evolution of Ghannouchi's Political Thought

Ghannouchi's work *al-Ḥurriyyāt al-'amma* is of particular significance to both its author and his followers. Ghannouchi mentions that on completing the first version in 1986, he was convinced that he had achieved a work of lasting importance, which "could overcome death itself": Islamic democracy.[4]

Ghannouchi argues strongly in his work that his notion of "Islamic democracy" is his own personal contribution to contemporary Tunisian Islamic thought. However, historical evidence with which to substantiate this claim is somewhat lacking. For example, there is no evidence of his using this term in his articles and speeches from the early seventies, when he was more interested in questions of education and morality rather than issues such as democracy and political freedom.[5] In fact, Islamic democracy was never mentioned in the context of his country's problems, and this was not because the Tunisian government at that time was especially democratic. As we have seen, Tunisia was then under one-party rule, which controlled all aspects of public life and was itself controlled by President Bourguiba.

The reason behind this curious silence was that democracy was actually not an issue for the Tunisian Islamists, at least not during the first years of their formation. Indeed, they began to pay attention to the concept only for pragmatic reasons. The first was as a result of the impact of the trade unions' demonstrations in 1978 and the consequent emergence of Prime Minister Mzali under the banner of openness and democracy. Second, by 1981 the Islamists had earmarked democracy as a possible ideological safe haven after the police's disclosure of their secret organisation on 5 December 1980. There was also the strong impact of the Iranian Revolution on the movement and especially on Ghannouchi, who saw it as the beginning of a new era that could bring into being an Islamic state that would "rule the world by the constitution of the Qur'an."[6]

These factors combined helped to change the Tunisian Islamists' message from being religious-educational to political, while as we have seen, the very idea of forming a political party came originally as an emergency measure to pre-empt a governmental attack. It was only once they had established themselves as a political party that the Islamists felt the need to produce literature that would address specific political questions, as Ghannouchi testifies:

> The issue of civil liberties in the Islamic state became my main topic of interest when the Tunisian Islamic movement began to change from the stage of propagating the principles of Islam in the face of a dominant foreign culture, to a different stage of a wider attachment to the needs of Tunisian society, and Arab society in general. That was more than ten years ago.

> At the top of these needs was—and still is—the issue of freedom. It was an intellectual necessity for the Islamic movement to offer clear answers to the challenges facing Islamic thought in a country such as Tunisia, which had become extremely Westernised. There was no other alternative.[7]

It was at this time that Ghannouchi began to read the relevant literature concerning the political issues in question,[8] and he used most of his time in prison between 1981 and 1984 in preparing the material for his book. On his release, he regained his position as *amīr* of the movement and was distracted for a time from writing up his findings. It was only when he was forced to stay at home under police surveillance that he finally managed to write the first draft of his book. After being released from prison once more in 1988 by President Ben Ali, Ghannouchi again concentrated on day-to-day politics rather than on formulating political theories. Only after going into exile in 1989 did he turn his attention fully to the concept of Islamic democracy. The book was printed finally in August 1993, with an important note from the author: he asserted that, in writing the book, his aim had not been merely intellectual; his objective had in fact been "an Islamic revolution that would uproot dictators from the land of Allah."[9] By this he was essentially referring to the "minority government" of President Ben Ali of Tunisia.[10]

This personal account of the emergence of Ghannouchi's book is important in that it reveals the political dimensions of his self-proclaimed, yet genuine interest in offering a democratic interpretation of Islam. It was clearly not an interest that had been of much concern at the time of the movement's beginnings in the late sixties, but was almost definitely a pragmatic response to the security and political concerns which the movement faced after 1981. Even though it became a political necessity to present the Islamic movement's views on democracy and political freedoms, these issues only really preoccupied Ghannouchi when he was in prison, under house arrest or in exile. Whenever he was in a position to play his traditional role as *amīr* of his movement, the democratic interest became secondary to the day-to-day running of the movement's affairs.

One should take into account, however, that Ghannouchi spent the greater part of the eighties and nineties either in prison, in hiding or in exile, which may help to explain why he was finally able to produce a lengthy, 382-page volume, showing a clear attempt to render his theory in an academic style, and with an extensive bibliography containing many classical, contemporary and even foreign-language references.

Islamic Democracy

If we may say that most Arab governments tend to favour secularising politics and society in their countries without actually implementing Western norms of democracy, the Tunisian Islamic movement seems to adopt quite the opposite position: it accepts Western democracy, with a few reservations, under the banner of Islamic *shūrā*, but still rejects secularism as a theory that separates religion and politics.

For Ghannouchi, there is nothing wrong with the institutions of Western democracy, such as elections, parliament, the rule of the majority, the multiparty system and freedom of the press. The main problem lies with "the political, national and material philosophies of the West, in that it separated body and soul, then ignored the soul, killed it, declared war against God and fought ferociously to put the human being in His place."[11] This, he explains, is why "secularism and, for example, nationalism, racism, the supremacy of values of profit, pleasure, dominance, power, utilitarianism, and the separation between religion and state are not fundamental parts of the democratic system."[12]

After making the distinction between democracy and secularism, Ghannouchi goes on to suggest that democracy may in fact be of Islamic origin. He speaks of "the way in which Western reason dealt with [the Muslim] heritage of engineering and mathematics, when it transformed them into concrete technology," and then invites Muslims to do likewise with Western democracy: "Why should we reject these political and industrial tools simply because they were made in the West? Is it not better to say that this is our merchandise brought back to us, and that wisdom is the aim of the Muslim and wherever the public interest is found then there is the *sharī'a* of Allah?"[13]

In this context, the real alternative for Muslims is to introduce what he defines as "Islamic" democracy:

> As the democratic system worked with Christian values and produced Christian democracies, and with the socialist philosophy to produce socialist democracies, and also with Jewish values to produce Jewish democracy, is it then impossible to work with Islamic values to produce Islamic democracy? We support this trend, and find in it a great relief, not only for the Islamic nation suffering from dictatorship, but for all humanity. [14]

According to Ghannouchi, there are two main pillars to this Islamic democracy: first is *sharī'a*, meaning the Qur'an and the *sunna*, *shūrā* is second, as the method by which Muslims can rule themselves. Allah is the "original governor" and the one with supreme authority over the universe, thus his ruling over human affairs reigns supreme over all other laws and ideologies. To uphold his rule is obligatory for every Muslim, and this is why Muslims should organise themselves politically and form an Islamic state. The Islamic state's *raison d'être* is the implementation of the *sharī'a*, which is also the source of its legitimacy; if it is not implemented, then it cannot command the obedience of the people. This is the framework within which the Muslim's obedience to his government becomes obedience to Allah, for which he will be rewarded. Conversely, his disobedience to the government becomes disobedience to Allah.[15]

In comparison with Western democracy, which Ghannouchi claims "kept searching for the basic values on which the state's juristic system should be built," he argues that the Islamic state has found in the texts of the *sharī'a* a solid base and a code of just laws; a canon not made by a ruling majority or a dominant class, but by Allah, the God of all. This canon is applied in detail by human institutions chosen by the people, wherein lies the authority of the *umma*, embodied in *shūrā*.[16]

After quoting those Qur'anic verses that emphasise the need to obey Allah's judgment in all disputes, Ghannouchi concludes that "there is in Islam a political system ordered by Allah, its rules as detailed in the Qur'an and *sunna*, and that obedience to them and their full acceptance is the line of demarcation between faith (*īmān*) and disbelief (*kufr*)."[17]

The Implications of Trusteeship (*Khilāfa*)

The wider framework for this "Islamic democracy" is a strong stand for the respect of human rights. In a speech given in Khartoum in 1979, Ghannouchi suggested in unequivocal terms that Islamists should fight against dictatorship in any form, even if it is practised by someone claiming to use it for the sake of Islam: "How can we tolerate it when Allah has forbidden it for his prophets? It is vital that we consider the human being as the trustee (*khalīfa*) of Allah on this earth and believe that *jihād* for freedom is a *jihād* for Islam." [18]

The trusteeship (*khilāfa*) of the human being is the cornerstone of the Islamic approach to human rights, according to Ghannouchi, for it is Allah who created humankind and decided that freedom and dignity should be the conditions for the human's trusteeship on earth. This notion is based on the Qur'anic verse: "We have honoured the sons of Adam; provided them with transport on land and sea; given them for sustenance things good and pure; and conferred on them special favours, above a great part of our creation" (17:70). The basic interpretation for this verse is as follows:

> The distinction and honour conferred by God on man are recounted in order to enforce the corresponding duties and responsibilities of man. He is raised to a position of honour above the brute creation; he has been granted talents by which he can transport himself from place to place by land, sea and now by air; all the means for the sustenance and growth of every part of his nature are provided by God; and his spiritual faculties (the greatest gift of God) raise him above the greater part of God's creation. Should he not then realise his noble destiny and prepare for his real life in the hereafter?[19]

Trusteeship is mentioned earlier in the second *sūra* of the Qur'an: "Behold, thy Lord said to the angels: 'I will create a vicegerent on earth'. They said: 'Wilt thou place therein one who will make mischief therein and shed blood? Whilst we do celebrate thy praises and glorify the holy (name)?' He said: 'I know what ye know not'"(2:30). The basic interpretation for this verse is the following:

> If man is to be endowed with emotions, those emotions could lead him to the highest [point] and drag him to the lowest. The power of will or choice would have to go with them in order that man might steer his own course. This power of will (when used correctly) to some extent gives him mastery over his own fortunes and over nature, thus bringing him nearer to the God-like nature, which has supreme mastery and will. We may suppose that the angels had no independent wills of their own: their perfection in other ways reflected God's perfection but could not raise them to the dignity of vicegerency. The perfect vicegerent is he who has the power of initiative himself, but whose independent action always reflects perfectly the will of his principal.[20]

Vicegerency for Ghannouchi is understood in the same broad sense, meaning that the human being

> had been honoured by his creator with a mind, will, freedom and the sending of the prophets to help him find the right path and to progress in the path of perfection, through his obedience to the *sharī'a* and the law of Allah, which, defined in its final version brought by the Arab Prophet Muḥammad (peace be upon him) the framework of the human being's life individually and collectively, leaving to him in this context wide domains, empty spaces, in which he was required to exercise his vicegerency by filling them [through his own *ijtihād*], thus making a combination between freedom and commitment, unity and diversity.[21]

The implications of Allah's special regard for the human being leads to the idea that the freedom and dignity of the human being in Islam are not only rights, they are also duties, as Ghannouchi argues:

> Rights here become holy duties which the human being is allowed neither to ignore nor be deprived of. He does not own these rights, as Allah is their only owner. The human being is merely a trustee of these rights, required to behave towards them according to the will of their owner … likewise the duty to reject slavery, and the opposition of dictators and the fight for freedom, justice, progress and the welfare of humanity. These are not simply rights but religious duties, which he will be praised for observing and punished for ignoring. It is in this way that human rights are deemed holy, as an Islamic concept, which makes it impossible for them to be denied or manipulated by a party, parliament or ruler.[22]

From the wide range of interpretations of the Qur'an, the view defended by Ghannouchi on the concept of trusteeship is one of many logical possibilities. If compared, for example with the views of the Egyptian Muḥammad Salīm al-'Awwā, a specialist in Islamic political systems, one may note that al-

'Awwā's studies do not base their analyses concerning freedom in Islam on the concept of trusteeship, but on other verses of the Qur'an, insisting for example on the freedom of belief and giving priority to rational arguments and evidence.[23]

The argument that it is impossible for Muslim governments or rulers to deny human rights because they are duties within the Islamic context is clearly questionable. First, there is the obvious problem stemming from the fact that Ghannouchi has attempted to interpret the Qur'anic text to match contemporary, and mainly Western, meanings of political freedom. For example, using this methodology, it would be impossible to find any acceptable political experience in Islamic history that meets the definition provided by Ghannouchi of "political freedom." It would be difficult even to find a similar experience in Western history, for democracy is a very recent product.

Further, even by classical definitions of "freedom" and "justice," it is clearly difficult not to see the numerous failures of successive Islamic states and dynasties in observing the essence of Islamic teachings with regard to running the affairs of political life, such as allowing for *shūrā* and tolerating different political opinions. Indeed, if the meaning of the concept of vicegerency had been as clear to Muslim rulers as it evidently is to Ghannouchi, Islamic political history might have taken an entirely different course. Ghannouchi himself admits this, but argues that Muslims and non-Muslims alike have lived under dictatorships, and that contemporary Muslim governments backed by the West are far worse than all previous Islamic régimes. He also adds that the deviation from a principle is not evidence of its denial, and that historically, Islamic states have been acknowledged by many objective, Western historians for their tolerance towards non-Muslims. However, the overriding answer, he asserts, is for Islamic thought to continue spreading its message of freedom and Islamic democracy, which is what Ghannouchi aimed to achieve with his book.[24]

Basic Human Rights in Islam

For Ghannouchi, the *sharī'a* is meant to serve the interests of the human being in both this life and the hereafter. He states that in classical jurisprudence, especially that of Abū Isḥāq al-Shāṭibī (b. 790 AH) in his famous book *al-Muwāfaqāt, sharī'a* is defined as safeguarding the five main interests of the human being: belief, life, mind, wealth and honour.

In this context, the starting-point for Islamic human rights is the freedom of belief. From the verse "There is no compulsion in religion"(2:256), Ghannouchi argues that compulsion is incompatible with Islam, and that religion depends on true faith and free will; every person must have the freedom either to believe in Allah and his teachings or to refute both.[25] Freedom of belief is thus the basis for the following rights:

1. Equality: all citizens of the Islamic state are equal before the law, both in their rights and duties, including non-Muslims.
2. Freedom to practice religious worship: Islam guarantees followers of all faiths the right to build places of worship and practise their religion there, provided they respect the religion of the majority.
3. Freedom to propagate non-Muslim religions, even if this includes arguments (though not abuse) to discredit Islam. In this instance, Ghannouchi mentions the example of the different prophets who engaged peacefully in serious dialogue with their opponents, and the historical debates that used to take place in the mosques and palaces of rulers between Muslim scholars and scholars from other religions.
4. Freedom and dignity of the human being: Islam not only stresses the rights of the human being in life, freedom and safety, but also, notes Ghannouchi, makes those rights sacred for both the individual and society, according to the fundamental principle of man's role as Allah's representative on earth. Aggression against men and women is forbidden to all. Human beings must be protected against hunger, illness, homelessness, spying and all manner of injustices, while dignity should be secured for all people, regardless of their colour, nationality or religion, and whether living or dead. This also implies the total rejection of all kinds of torture and all kinds of illegal interference in the private lives of people.
5. Freedom of thought and expression: Allah has equipped the human being to be his representative on earth by giving him reason, will and freedom, leaving him to choose his path and assume his responsibilities. Islam makes "thinking" a duty, because through it vital issues are decided, such as the question of belief and its consequences in life.
6. Freedom of private ownership: Islam secures the right of private ownership in the context of the human being's trusteeship on earth, but Allah in fact owns everything. This implies that private ownership must serve the whole society, not just the selfish goals of the individual. One of the benefits of allowing private ownership, Ghannouchi argues, is to distribute power between people through the sharing of wealth, which gives democracy its social aspect and prevents the state or an élite class from controlling material wealth, and thus, political power.
7. Social rights: Ghannouchi includes in this respect the rights to employment, health care and social security. He quotes the Egyptian writer 'Abbās Maḥmūd al-'Aqqād and his summary of Islamic social democracy: "Islam rejects all kind of profiteering and considers as holy and sacred the rights of employment. In fact, democracy does not need more than these two principles to be stable and successful." Ghannouchi argues that it is a vital duty for the government to find employment for people and secure the rights of workers against any form of profiteering by their employers.

> Health care is also a fundamental aspect of human rights since it is linked to the essential right to and sanctity of life. The rights to have a family life and to education are also among Islamic social rights.[26]

In total, these rights represent the framework within which Ghannouchi presents what he sees as a modern Islamic political régime. It appears that there are no major differences between his views on this issue and the general Western approach to human rights, except in the religious basis he employs.

The People's Authority

The projected image of Ghannouchi's Islamic régime is that of a presidential system, with an elected president and parliament, where the *sharīʿa* is the supreme authority over the constitution and all laws. It is also a missionary system in the sense that the government must look after Islam and its promotion as the state religion, as well as keeping a vigilant eye on every action or movement that endangers the state, and to deal with it using the appropriate measures.[27]

If the *sharīʿa* is the main pillar for any Islamic government, Ghannouchi also argues that *shūrā* is the second most important factor to give a régime its true identity. He refers to verses such as: "O ye who believe, obey God, and obey the Apostle, and those charged with authority among you. If you differ in anything among yourselves, refer it to God and his Apostle, if ye do believe in God and the last day"(4:59); and: "Consult them in affairs (of moment)" (3:159); and: "Who (conduct) their affairs by mutual consultation" (42:38). Ghannouchi concluded from these verses that *shūrā* is thus a symbol for both the Islamic state and the Islamic nation, and that it is not simply a secondary religious ruling, argued for in only two verses of the whole *sura*. Rather, "it is in fact a fundamental pillar of the pillars of religion and a necessary result of trusteeship, meaning that the power of Allah is given to the people."[28]

Verse 4:59 is the basis for the Islamic political, social and religious order, according to Ghannouchi, because it states the supremacy of the authority of the Qur'an and the *sunna*, as well as stating the authority of the people, within the boundaries of the *sharīʿa*: "the Qur'an and the *sunna* are the ultimate law that governs the behaviour of rulers as well as all Muslims. They are also above any invented law [arrived at] through *ijtihād*, meaning that it must not contradict any rule of the Qur'an and that it must serve the main interests of the *sharīʿa*."[29]

Ghannouchi argues none the less that the *umma* is the source of legislation:

> Although the prime source of legislation in Islam is Allah's will, as is reflected in revelatory text from the Qur'an and the *sunna*, the *umma* should actively participate in legislating. The reason for this is that, making the final *sharīʿa* eternal re-

> quired limiting the text of the revelation to legislate only the main principles ruling human relations and not to elaborate on details and minor issues, except for in a few cases such as legislating for the punishment of a major crime and for certain issues related to the family; legislation that helps form the overall shape of Islamic society. This means leaving the details of this shape to the legislative effort of the *umma*, which changes with the times, and it is a respectable endeavour, for the *ijmāʿ* of the *umma* is considered one of the [religious] sources of legislation.[30]

After discussing the various ways in which the authority of the people may be expressed, Ghannouchi concludes by describing two forms of representation. The first is a direct form by referenda and general elections concerning major and vital political issues, such as entering military alliances, the direction of the state's main policies and the choosing of a leader. In issues such as these, there is no alternative but to seek the people's verdict by direct consultation. This direct *shūrā*, says Ghannouchi, is the textual implementation of the Qur'anic teaching which calls for the participation of all people in making the general policies of the state.

Second is an indirect form of representation, by an elected body of "those charged with authority among you," being people of good Islamic conduct who would form a committee of *shūrā* (parliament) that would play the role of control and guidance for both the government and the people. It would also make policies and laws within the framework of the *sharīʿa*. Non-Muslims could be elected to this body, and prominent experts in fields of specific importance would be assigned to it, along with the elected members.[31] Ghannouchi does not specify, however, how these "men of expertise" would be chosen and by whom, but it may be assumed that this would be the right of the elected members of the *shūrā* council, as is the case with *al-Nahḍa* council.

Ghannouchi insists that these two forms of consultation must go hand in hand with a committee of prominent religious and legal scholars with the main responsibility of making sure that all laws passed by the parliament or the government were compatible with the rules of the *sharīʿa*. This body could be considered a vehicle for collective *ijtihād* on all matters related to religion; Ghannouchi has compared it to the high court of the United States, the state council of France or the committee for safeguarding the constitution in Iran.[32] Again, he fails to specify the criteria on which the members of this body would be selected, and whether they would be elected or nominated.

One of the main duties of the *shūrā* council would be to nominate one or more candidates for the post of president of the Islamic state, without closing the door for other candidates not nominated by the council. The people would then be called on to elect their president, who would oversee the implementation of the *sharīʿa* and take care of the nation's interests. Having the president voted into office by the people, Ghannouchi argues, would ensure a popular mandate for the president, and free him from being tied to the narrow political calculations of the élites and factions assembled in the parliament. The president could then concentrate on serving those who selected him.[33] This is one of the reasons Ghannouchi puts forward for favouring a presidential political system; he also mentions the fact that the candidate will be chosen first by "those charged with authority among the population," guaranteeing a better relationship between the legislative and executive powers, and a better understanding between these powers and the public.[34]

There are, however, a number of necessary conditions that should be met by any candidate for the post of president of the Islamic state. Ghannouchi summarises them below:

> The president of the Islamic state must be a Muslim, a good Muslim. The minimum in this regard is for him to be known to have the correct religious beliefs and conceptions, to care for religion, to like knowledge and scholars, to observe his religious duties and abstain from forbidden deeds, to be honest and of a good character with a strong personality and physical capability, thus being able to serve the nation and lead it properly.
>
> He must be of an age of maturity and stability (40 years), having known life and the people and being known by them, being satisfactorily concerned with the needs of the people, and ensuring the testimony of those charged with authority among the population that he has religion, honesty and competence, and that he is the best among them. It would also be a bonus for him to have a transparent soul and enlightened thought [in religious terms], thus being able to play the role of the educating shaykh and the leading reformist [again in religious terms]. He must also have the consensus approval of the nation or its majority.[35]

It should be mentioned that some of the conditions listed above would be rather difficult to "prove" in a candidate since they are so subjective. However, the overriding emphasis is on the religious aspect because the main duty of the president is to uphold the *sharī'a*. It may also be seen to reflect the yearnings of many Muslims for a "true, honest" Muslim ruler, for the majority of rulers in both Islamic and modern history have essentially been politicians playing the power game both within and outside the boundaries of the *sharī'a*. At the same time, these conditions give Ghannouchi himself an edge over his political rivals in Tunisia, where he is seen as the one politician whose agenda focuses on religion and religious duties.

Ghannouchi's book goes into enormous detail about the classically defined duties of the president and his rights: such as his duty to liaise with the ordinary people, to be humble and faithful; and his right to be obeyed by the nation for as long as he obeys the rules of the *sharī'a*; and to encourage "those hearts that do not like him to do so, and to ensure people's love for him in the interests of the whole nation."[36] This quotation was used by Ghannouchi in detailing the president's rights, also outlining another nine rights of this nature, all deriving from the special religious status of the leader of the Islamic state.

Thus, in short, the Islamic concept of *shūrā* is basically expressed in electing a parliament, and a president who is initially nominated as a candidate by the parliament. The two main constitutional questions that arise here are: what is the relationship between the legislative and executive powers, and what is the relationship between these powers and the committee of leading religious and legal scholars, suggested by Ghannouchi to supervise all the laws in the land?

In response to the first question, Ghannouchi's answer is imprecise. He mentions, rather vaguely, that there are many options available to Muslims according to their own specific circumstances and at any given time, and argues that there is nothing in Islam which gives a specific ruling about the relationship between the legislative and executive powers. However, he does provide a sufficient number of hints to suggest that he disapproves of the rigid separation of the two, as an idea deriving from the Western culture of "a conflict of interests," which has turned Western democracy into a simple mechanism with which to institutionalise this conflict and keep it in check.[37]

Ghannouchi also argues that this notion of rigid separation is not compatible with Islamic concepts and values: it is contradictory to the *bay'a*, which the body of *shūrā* gives to the *amīr* in order to listen to and obey him, and would give the ultimate power of authority to the legislative council rather than to the *amīr* or head of state, thus making him accountable to this council and not to the people who voted him president. This may be acceptable only if the head of the Muslim state were to hold a merely nominal or protocolic position, a Western system of which no scholar of Islamic politics has ever approved, nor which has been approved by the revelation or any other Islamic ruling, or by the experiences of the first caliphs.[38]

The alternative, according to Ghannouchi, is the principle of co-operation (*ta'āwun*) between the different powers, to ensure that the word of Allah prevails.[39] The judiciary should rule independently over all individuals including the president, while legislative power is that of the Qur'an, the *sunna* and the supervising committee of scholars. This would mean that the Islamic state is essentially an executive state based on both the parliament and the president's authorities.

Ghannouchi then goes on to address the second, and more important, question:

> In the case of a dispute which cannot be solved between the president and the *shūrā* council of which he is president [implying that Ghannouchi favours the president himself being the president of parliament], the matter should be examined: if those opposed to the president represent a consensus (*ijmā'*) or something close to it, such as a two-thirds majority, then the president should bow to this consensus because it is one of the pillars of Islam. If it is a small majority, then the issue should be put before a committee of experts if related to a matter of legislation, and its ruling should be obeyed by all. If the issue of conflict is related to general policy, then the president may have the right to veto the council's decision and refuse to execute it.[40]

Ghannouchi also provides another option for the president in the face of parliamentary consensus: if he chooses not to accept the majority decision, he can take the matter to the people in a referendum. Whoever loses the vote, be it the president or the parliament, must then concede their position.[41]

Ghannouchi clearly favours a presidential system with strong powers. His view that the president may also preside over the *shūrā* council is a clear sign of his rejection of the notion of separating the executive and the legislative powers. However, this may in fact throw all of his book's democratic credentials into question, for it is clear that as president of the parliament itself, this allows the president too much unchecked authority to be truly "democratic."

His defence, which is his argument that real legislative authority lies with the *sharīʿa* and the committee of scholars, provides more questions than answers. For example, who will nominate these scholars: the president, the parliament or the people? Is their responsibility limited to giving their opinion on laws passed by the parliament or by the government, or does it include making new laws independent of the parliament or the government? Despite the importance of these questions, Ghannouchi fails to provide an adequate answer to the first, and merely offers confusing hints regarding the second.

In fact, there is no discussion of how the members of the committee of prominent scholars and experts of law should be chosen. One should assume, therefore, that it is the right of the president or parliament to select them, which opens the door for either of these powers to be politically manipulative, in order to bring as many "friendly" scholars as possible on to this important committee. It may therefore also be safe to assume that scholars known to hold views opposing those of the ruling class will not be nominated for this prestigious position. In short, this is a highly sensitive area, with wide-ranging possibilities for disagreement between the scholars on a number of issues.

Ghannouchi does emphasise, however, that the main responsibility of this body would be to make sure that the laws passed by the parliament or the government were compatible with the *sharīʿa*. He also mentions that it would be responsible for the "orientation" of both the government and public opinion, by "single and collective *ijtihād* on issues related to the *sharīʿa*, without being binding [to the state]."[42] This role of "orientation" and of creating new laws should initially be entrusted primarily with the parliament, a view that Ghannouchi reiterates throughout the book.[43] Within this context he argues that in religious terms, the *umma* should be the source of legislation, because the *sharīʿa* has left a wide, empty space for the changing of *ijtihād*, as exercised by Muslim public opinion and organised by the body of the *shūrā* council.[44]

However, he also states, when discussing the relationship between the powers in the Islamic state, that the responsibility of legislation should normally be left to the Qur'an, the *sunna* and the supervising committee of scholars, and that the *shūrā* council is in fact a part of the executive power, presided over by the head of state, in an Islamic state of an executive nature.[45]

These conflicting statements seem to reflect the difficulties Ghannouchi faced in trying to set up a detailed, modern Islamic political system that would be compatible with both classical Islamic models and the norms of modern Western democracy. One of the major consequences of giving a wide range of powers to the supervising committee of scholars is that it increases fears of an autocratic state run by religious scholars, accountable only to what they see as the divine truth of Allah. Ghannouchi and his colleagues have always distanced themselves from this concept. At their press conference of 1981, they assured the public:

> Let us make it clear that Islamic thought and the MTI reject the idea of an Islamic autocratic government, because there is no one who can claim to be the official spokesman of Allah or Islam, and thus oblige people to follow their own interpretation of Islam as the sole legitimate interpretation. Rather, we will propose our own policies and views and will not oblige anyone to accept them. We expect the same from other parties and recognise their rights to deal seriously and responsibly with Islam. While we reject theocracy, we do believe in *shūrā*; we will propose our programme and accept the people's verdict, for or against us.[46]

If this does not represent a real safeguard against a dictatorial autocratic state, *al-Nahḍa* in general and Ghannouchi in particular have spoken of a number of assurances or "guarantees against dictatorship." These include the right for all to form political parties; their 1981 declaration stated the movement's rejection of the one-party system, and their statements to the press were unequivocal:

> Islam says that there should be no compulsion in religion. Our problem is that others want to impose their policies on us by force. Let us make it clear: we have never opposed the existence of any political movement, even if it is totally against our views, including the communist party. This is a fundamental Islamic position, because it is for the people, not us, to choose this party or any other for government. Indeed we have proven this through our relationship with other opposition parties.[47]

Ghannouchi argues that, in an Islamic state, the multi-party system is the best—if not the only—way to strengthen society *vis-à-vis* the state, especially when the state deviates from the path of justice. What is more, he argues that political parties have a great role to play in educating people about their rights and duties and spreading values throughout society, while part of their mission is also to nurture a wide range of well-educated politicians and potential leaders to serve society. Political parties should not need official permission from the government to operate, because their activities are in fact one form of executing Allah's order to "enjoin what is right and forbid what is wrong."[48]

In Ghannouchi's view, the independence of the judiciary and freedom of the press are other guarantees against dictatorship, arguing that in Islam justice reigns supreme over all people, including the president and the government. In order to ensure this supremacy, there should be no governmental participation whatsoever in judicial affairs.[49] Similarly, a free press constitutes a means by which the people can keep an eye on every aspect of public affairs, and put a stop to the wrongdoings of the government.[50]

International charters are also included in the list of Ghannouchi's guarantees against dictatorship. He argues that there is nothing in Islam that opposes the building of an international order between peoples and civilisations based upon justice and equality. An example he gives is of a charter opposing aggression and injustice around the world.[51]

The Status of Non-Muslims

Another important issue facing all advocates of an Islamic state is the status of non-Muslims and the extent of their rights within a missionary state that sees itself as a servant to the values and teachings of Islam. Ghannouchi provides his own analysis both in *al-Ḥurriyyāt al-'amma fī al-Dawla al-Islāmiyya*, and in another book, *Ḥuqūq al-Muwā ana* [The Rights of Citizenship],[52] entirely dedicated to delineating the rights of non-Muslims in the Islamic state.

The starting-point for Ghannouchi's thoughts on the matter is the contract of *dhimma* used in classical Islamic law schools to organise the residence of non-Muslims in the Islamic state, which he sees as the equivalent of today's laws of citizenship. He has adopted this equivalence from another contemporary Muslim writer, 'Abd al-Karīm Zaydān, and also refers to Mawdūdī's definition of *ahl al-dhimma*, or

> all those non-Muslims living within the boundaries of the Islamic state and who accepted to give it their allegiance and to obey its laws, whether they were born in it or migrated to it from abroad and asked the government to consider them among the people of *dhimma*. Islam offered these people, in the internal laws of the country, the same rights offered to Muslims, on equal terms.[53]

This is similar in essence to what a Muslim must do nowadays if he or she wishes to become a citizen of a European country, for example.

To go into further detail: Ghannouchi argues that non-Muslim citizens of the Islamic state may enjoy the same basic human and political rights guaranteed to Muslim citizens. This includes the right to equality before the law, freedom of thought, belief and expression, including that of defending their religious beliefs.[54] Non-Muslim citizens of the Islamic state are also free to form their own political parties as long as they recognise the sovereignty of the *sharī'a* and pledge their loyalty to the state. They may also invite Muslims to discuss their respective religions provided that they honour the general norms of dialogue; this is not seen to represent a "danger" to Islam, which cherishes intellectual dialogue with all faiths and ideologies.[55]

Needless to say, however, there are certain exceptions for the non-Muslim, which, Ghannouchi argues, are none the less fair and just:

> His citizenship remains special and only becomes complete on converting to Islam. This means that he may continue to enjoy freedoms not allowed to the Muslim in his personal life, regarding his food, drink and marriage, but also that he is deprived of certain rights enjoyed by a Muslim, such as serving in key political posts that influence the nature of the state [such as the presidency]. On the other hand, the non-Muslim citizen remains free of certain requirements imposed on a Muslim, such as avoiding certain forbidden deeds [drinking wine, for instance]. There are a limited number of exceptions that do not affect the principle of equality, a highly respected principle in the Islamic state.[56]

In other words, Ghannouchi argues that true equality suggests that the religious nature of the Islamic state should not be undermined, and that non-Muslims should recognise the fact that the main orientation of the state is Islamic and its main duty is to uphold the *sharī'a*. This explains why only Muslims may fill key posts, such as the presidency, the speaker, the council of *shūrā*, the chief of the army or the membership of the supreme council of justice. Although non-Muslim citizens are free to choose their beliefs, at the same time they must recognise the right of Islam, as the majority religion, to organise and direct public life. Similarly, their political parties should not aim to undermine Islam's leading position.[57] While not eligible to serve in key posts, non-Muslims may still serve in the Islamic state as members of parliament or even as ministers; the main guideline here is that they are deprived only of taking up posts that have a direct relationship with the state religion.[58]

There remains the question of whether the non-Muslim citizen should pay a special tax to the Islamic state, known in Islamic history as the *jizya*. Ghannouchi's response is that, historically, the *jizya* was imposed by the Islamic state as a contribution towards ensuring the security of the country, because non-Muslims were not enlisted into the army. In this sense, it was a form of defence tax that was not imposed on Muslims because they were required to join the army and defend their state. Consequently, he concludes that this tax may be abolished for all non-Muslim citizens who accept doing military training and are ready to join the army if required to defend the country.[59]

With regard to Islamic *ḥudūd*, as carried out for certain major crimes, Ghannouchi tends to argue for the application of Islamic criminal law to all citizens of the Islamic state where possible. This, however, should not ignore the need to accommodate different circumstances and allow for different laws within the Islamic state when there is a non-Muslim majority in some of its provinces.[60]

Finally, here is how Ghannouchi summarises the nature of an Islamic state:

> The Islamic government is one in which:
>
> 1) Supreme legislative authority is with the *sharī'a*, that is the revealed law of Islam, which transcends all laws. Within this context, it is the responsibility of scholars to deduce detailed laws and regulations to be used as guidelines by judges. The head of the Islamic state is the leader of the executive body entrusted with the responsibility of implementing such laws and regulations.
>
> 2) Political powers belong to the community (*umma*), which should adopt a form of *shūrā*, which is a system of mandatory consultation.[61]

This model of government, which Ghannouchi insists is a religious duty for all Muslims to establish, must work for the welfare of the people within the general framework of the aims of the *sharī'a* as defined by al-Shāṭibī. He argues as follows:

> The general inclination in contemporary Islamic political thought is to adopt al-Shāṭibī's concept of the purposes of religion, namely that religion was revealed only to fulfil and protect the needs and interests of mankind in this life and the hereafter, and as a general framework for tackling new problems in Muslim society. Within the framework of this general perspective or purpose of the *sharī'a*, the details of religion find their appropriate place as branches of the fundamentals. Within this same perspective, all new problems in the lives of not only Muslims but of all humanity, can find proper solutions that guarantee the fulfilment of their requirements. [62]

Within such a wide framework, the proposed Islamic government may opt for a broad range of policies in all areas of social life, as long as they do not violate the explicit laws of the *sharī'a*. Indeed this framework is so wide that the government may effectively disregard those laws in certain circumstances, as Ghannouchi illustrates:

> Some of these principles [of the *sharī'a*] apply to standard situations and normal circumstances, while others deal with exceptional situations and extraordinary circumstances. An example of the latter is the principle of "necessities eliminate prohibitions" (necessity knows no law). This principle is Qur'anic. Allah says: "But if one is forced by necessity, without wilful disobedience nor transgressing due limits, then he is guiltless" (2:173).
>
> Similar to this are the principles of balancing between the better and the worse and opting for that which seems to best serve the general interests of the people, and the principle of outcomes or consequences, according to which acts are judged on the basis of what they lead to. The decisive criteria in all cases are the fulfilment of the needs of humans and serving their best interests.[63]

In classical terms, Ghannouchi's views do not reflect those of one single traditional school of Islamic law (*madhhab*); like most contemporary Islamist writers, he examines different interpretations of the *sharī'a* and selects what he thinks is best in responding to contemporary circumstances. He is not a *mujtahid* inventing new views or exploring new avenues in Islamic jurisprudence, but a reformist, or a politician who is anxious to clarify his position on important political issues. As we have seen, most of his preferences reflect a strong desire to present an Islamic political system that is compatible with contemporary norms of Western democracy.

Despite its ostensibly "moderate" and "modernist" appeal, *al-Nahḍa*'s political approach to the issue of civil liberties is still the source of many doubts among Tunisians. For example, in the first three years of President Ben Ali's rule, there were many discussions focusing on the question: "is the new strategy of the Tunisian MTI [i.e., its strong support for democracy] mere manipulation, to win the support of public opinion for their demand for legal recognition, or is it a reflection of the roots of the movement in civilian society and Tunisian political culture?"[64]

The Tunisian sociologist, Dr Abdalkadir al-Zghal, who openly raised this question, has been rather lenient by Tunisian standards: he notes that Ghannouchi is a politician who wishes above all to please his supporters, even to the extent of expressing different views to different audiences among Tunisia's pluralistic society. None the less, he remarks that this should not be deemed unusual, for Ghannouchi is not a "theologian like Ibn Taymiyya, ... but a politician who tries to expand his political base and reduce opposition among his partners and opponents."[65]

No doubt this interpretation does not please Ghannouchi, for, after all, he presents his views in religious terminology and as an Islamic thinker first, and as a political leader second. Evidence suggests, however, that political pragmatism is not always compatible with providing religious answers to important issues. Striking a balance between the requirements of religion and those of political tactics and manœuvres is not always easy, to which fact, no doubt, Ghannouchi can testify.

However, an analysis of his words and his deeds during his political career in the mid-1990s indicates that his political considerations have been very much stronger than the religious. To many of his detractors, the stance Ghannouchi took during the second Gulf War against the Western military presence made him one of the Iraqi president's strongest supporters, despite the fact that he was not alone in his support for President Ben Ali and the majority of Tunisian politicians also sided with Iraq. However, the special criticism aimed at Ghannouchi in this instance concerned his total disregard for Saddam Hussein's undemocratic record, and the extent of the enthusiasm he expressed for the Iraqi leader. Ghannouchi praised his wise and courageous leadership and declared that the man "who dared to attack Israel and resist in the face of united international aggression deserved to give orders and to be obeyed [by Muslims], and to have the right to money and lives,"[66] which suggests that Hussein had the right to demand money and volunteers from the Islamic nation because he was the leader of an "Islamic" battle.

The apparent ease with which Ghannouchi portrayed Hussein as an "Islamic leader" was a worrying fact, even to his close Islamist friends, such as the Egyptian Fahmī Huwaydī, who wrote:

> The record of the Iraqi régime, which the statement [of Ghannouchi] asked us to support, is full of pages of atrocities and dictatorship. The battles it has fought during the last twenty-five years were first against the Arab people of Iraq, then against the Muslim Kurds, then against the Islamic Republic of Iran, then with the Lebanese Maronite General Aoun, and then against the people of Kuwait ... How is it that Professor Ghannouchi did not hear in his exile about the tens of thousands of victims of that régime, or about what is left of them after the killing, torture and poisoning? [67]

It was not long before pressure from some of Ghannouchi's colleagues, particularly following the failure of the movement's plan to "enforce liberties" in Tunisia, forced him to change his opinion to the extent that he began to criticise Hussein's democratic record and declare: "I cannot accept that either myself or the Islamic nation should be led by Saddam Hussein and others like him who are neither democrats as I understand the term nor Islamists as I understand Islamism."[68] He then went on to recommend Hussein's resignation from power.[69] This was basically a political about-turn, undertaken in an attempt to win back influence and support for *al-Nahḍa* from Saudi circles in particular, although it proved to be a futile gesture. One London-based Saudi newspaper did not even bother to publish a later statement issued by Ghannouchi in which he condemned Iraq, dismissing it as a political manœuvre by which to seek support for his struggling movement.

A second reason for questioning the democratic claims of the Tunisian Islamists arose as revelations surfaced about their secret military wing, which they had set up to support their political struggle. As we have seen in earlier chapters, the nature of this wing meant that it was forced into secrecy, not only with regard to the authorities and other political parties, but also with regard to the vast majority of *al-Nahḍa* members and supporters. Indeed it was only after the aborted coup of November 1987 that most people came to know of the military wing's existence. The problem it posed was that it forced the assumption that the balance of power lay with whoever in the movement was in control of its military section, a scenario not entirely compatible with the democratic norms as suggested by Ghannouchi.

Following the disastrous results of *al-Nahḍa*'s confrontation with the régime, many supporters of the movement blamed Ghannouchi for forcing it into a political battlefield. Long before Mourou's resignation from his post as secretary-general, there had been a great deal of talk about whether Ghannouchi's statements were truly representative of *al-Nahḍa*'s true position, and about the legitimacy of his status as president of the movement. The official view expressed at the time by Ali Laridh was that Ghannouchi did in fact represent the movement whenever he signed a statement on *al-Nahḍa*'s behalf.[70]

This accusation later gained even more momentum, notably following the resignations of Mourou and his sympathisers. When asked at that time what advice he would like to give to Ghannouchi, Mourou advised him to abandon the use of violence in the domain of Islamic political action.[71] He argued further that Ghannouchi could not be accepted as a member of the breakaway party he intended to form, for, in the eyes of public opinion, this would merely be seen as double standards.[72] Even to a close personal friend of Ghannouchi and founder member of *al-Nahḍa* such as Abdalkadir al-Jadidi, Ghannouchi's statements in exile were not representative of the movement's views and "embarrassing even to his friends, unthinkable to intelligent minds, and annoying for the intellectual and man of trust."[73]

There were other, less prominent supporters of *al-Nahḍa* who also accused Ghannouchi of manipulating the movement according to his personal views and plans.[74] Ghannouchi's defence was always that he had remained faithful to the internal democratic rule of his organisation, and that the complaints of these Islamist dissidents did not represent an "intellectual *ijtihād* or a new school of thought."[75] For his critics, this dismissal of their views was further proof of Ghannouchi's undemocratic tendencies. None the less, he chose not to take their complaints seriously and did not feel obliged to resign from his post as declared president of *al-Nahḍa* in exile.

A Note on Tunisian Modernism

Most of these doubts and accusations, both from within *al-Nahḍa* itself and from outside observers and opponents, came to the surface during the mid-1990s largely owing to the movement's double defeat: in Tunisia as a result of a government-organised campaign, and internationally following the defeat of Saddam Hussein.

This should not, however, undermine the gains that the movement made in presenting itself as a democratic Islamic trend. This approach in fact has been appreciated by many, to the point that al-Zghal has argued that Ghannouchi and Bourguiba are both "above all the products of Tunisian civil society, well entrenched in the long tradition of exchange between the two shores of the Mediterranean and continuing the reformist tradition of political men such as Khayr al-Din and [members of] the *'ulamā'*, such as Taher ben Achour."[76] In this sense, al-Zghal ranks Ghannouchi alongside Tunisia's leading modernists, a point that warrants closer examination.

From the time of the rule of Ahmad Bey (1837–55) to that of Bourguiba, Tunisia has seemed more prepared than most other Islamic countries to adopt

Western-style reforms within its political and social structures. Slavery was abolished in 1846, equality between Muslims and non-Muslims before the law was granted in 1857 by Mohamed al-Sadik Bey in a document known as the "Pledge of Safety" (*'ahd al-amān*), which also granted foreigners the right to acquire properties in Tunisia. A new constitution was adopted in 1861, giving more powers to a consultative council and to ministers at the expense of the Bey. This period also witnessed the rise of the man whom many were later to call the "father of the Tunisian renaissance," Khayr al-Dīn.[77]

Khayr al-Dīn was Tunisian prime minister from 1873 to 1877, and was most famous for his book *Aqwam al-masālik fī ma'rifat aḥwāl al-mamālik* [The Straightest Way to Know the Conditions of Kingdoms], first published in Tunis in 1867. The importance of the book lies in its introduction, in which Khayr al-Dīn specified his two main purposes:

> First, to urge those who are zealous and resolute among statesmen and men of religion to adopt, as far as they are able, whatever is conducive to the welfare of the Islamic community and the development of its civilisation, such as the expansion of the boundaries of science and learning, preparation of the paths which lead to wealth from agriculture and commerce, the creation of industries and the eradication of reasons for unemployment. The basis for all this is good government, from which result security, hope and good work, as seen in the European kingdoms.
>
> Secondly, to warn those who are heedless among Muslims about persisting in closing their eyes to what is praiseworthy in the practice of adherents to other religions, but which is in conformity with our religion, simply because they have the idea engraved in their minds that all acts and institutions of non-Muslims should be avoided, and that their books on these issues should be denounced and not mentioned. They strongly criticise whoever appreciates them, despite the fact that this behaviour is clearly not correct.[78]

Khayr al-Dīn also called for governmental reforms, to ensure wider consultation with eminent leaders in society, and suggested many ideas for solving the problems of Tunisia's stagnant economy. What is more, he insisted on the need for benefiting from Western methods and policies in running Tunisia's public affairs, and on the role of the state in leading the way to adopt these necessary changes and reforms.

There are two main remarks that may be made about the Tunisian renaissance: the first is that it was strongly influenced by Western, and especially French, domination. The constitution of 1861 was in fact written by the French, arabised by the Bey's secretary and approved by Napoleon III. Khayr al-Dīn himself was clearly influenced by Western policies and was intent on achieving success by Western standards, although not at the expense of the *sharī'a* and the Islamic alliance with the Ottoman empire. Bourguiba's situation was clearer still from his strong cultural and political links with France. A second, but equally important, remark is that all reformists and political leaders in contemporary Tunisia have emphasised the role of the state as a key factor in carrying out their policies. Discussing the changes that occurred in Tunisia in the period leading up to 1881, Michael Brett has described this period as that of "the great increase in direct state control of the population," leading to a state of affairs where 'the traditional social hierarchy was severely weakened, its place taken by a general levelling ... of the population."[79]

This trend helped to form a new class that the state could control and manipulate. According to Brett:

> The word "class" is too dangerous to be used without qualification. Nevertheless it seems an appropriate term for a rural population whose traditional social and economic structures had been subordinated to the fiscal pressure of the central government, creating a homogeneous class of taxpayers and subjects, and under Ahmad Bey, of conscripts to the army, whose new condition increased steadily in importance at the expense of the old. "Class" likewise seems appropriate for those who undoubtedly benefited from this new direction taken by the state, its agents who profited from its patronage.[80]

By 1987, Bourguiba's rule had all but accomplished this very significant social transformation; the state had become the only and absolute source of power in Tunisia, at the expense of all other traditional forms of alliance, or even modern alliances such as syndicates or political parties. Thus when President Ben Ali assumed power, he had only to crush the Islamic movement to reclaim total control over the country, with state power similar to or even stronger than that of communist régimes in the Soviet Union and Eastern Europe.

For this reason, it is hard to agree with those such as al-Zghal and others who speak of "Tunisian civil society and its political culture," as though democracy was a dominant component of Tunisian political life throughout the nineteenth and twentieth centuries. Although Tunisia may rightly deserve special mention on account of its Western-influenced, modernising trends, this must exclude democratisation. In fact, like many Muslim governments, it has often chosen to disregard democracy and the will of the people in order to enforce Western-oriented reforms.

Because Tunisian modernism was a state-owned and essentially undemocratic project, the efforts of *al-Nahḍa* to present itself as a democratic party failed to result in any genuine acceptance on the part of the country's ruling and influential élites. Rather, these élites merely viewed the Islamic movement as a dangerous threat to all of the Western-oriented reforms that had been made throughout the previous century, which were of far more significance than the possible positive aspects of a legal Islamic party. Indeed, these élites were unable even to accept differences within their own camp, let alone accept those who wanted to bring about an Islamic state.

Another explanation may be found in the theory of the Tunisian writer Saleh Bashir, in which he argues that Tunisian modernism, which was much more successful than the limited achievements of most Arab-Islamic countries, really evolved under dictatorship, not democracy. He also argues that because modernism managed to strengthen its roots in all classes of Tunisian society, it became a source of legitimacy that every ruling élite aspired to monopolise, in the same manner as religion or the will of the nation. Thus by claiming to be modernists, these élites presented their form of dictatorship as a legitimate defence of modernism.[81]

The Response of Tunisian Élites

As we have seen, Ghannouchi's book on civil liberties in the Islamic state may not necessarily be seen to represent the "official" view regarding the political theory of *al-Nahḍa*. Among his works, however, it is the most important document in the movement's political literature, which provides a detailed theory of the Tunisian Islamists' Islamic state. It is also true that this proposal for an Islamic state is compatible with the general line expressed in *al-Nahḍa*'s official statements issued from the date of their formation as a political party in 1981. A few points emerge when one looks closely at the various Tunisian responses to this position.

Ghannouchi's book was published in 1993, by which time *al-Nahḍa* had been banned in Tunisia. As a result, the book was not distributed in the country and there was no opportunity to review or discuss its ideas in the national press, although *al-Nahḍa*'s ostensibly liberal views were well known, for much of the movement's political rhetoric concerned democracy and the multi-party system. None the less the Tunisian government, under both President Bourguiba and his successor Ben Ali, chose to ignore *al-Nahḍa*'s claims and focus on fighting it as an extremist political party using religion for its partisan aims. The official view was expressed in the following quotation, taken from President Ben Ali's speech at an official symposium organised with the United Nations Educational, Scientific and Cultural Organisation (UNESCO) in 1995 on the subject of "The Teaching of Tolerance in the Mediterranean Basin":

> We are engaged in a fight against extremism which has shown by its terrorist methods not only hostility to the values of tolerance but also to the spirit and precepts of religion. This extremism seeks to wipe out the achievements of modernity, to encourage hatred and fanaticism, and to destroy the universal values and norms which govern human rights and all civilisations and religions.[82]

This statement reflected the Tunisian government's outright rejection of the Islamists' political activities, and confirmed that opportunities for the discussion of views and ideas between the two sides seemed to be slender, at least for the present.

Few Tunisian writers and analysts have publicly written positively about *al-Nahḍa*'s "democratic message." Together with al-Zghal, Elbaki Hermassi is possibly the most prominent personality among Tunisian academics to have noted this, in as early as 1984, in his paper to a conference about the contemporary Islamic movement in the Arab world. In this paper he spoke of a new political approach taken by the Tunisian Islamic movement since 1980, one "that reversed the traditional equation 'Allah, the ruler, the people' to one of 'Allah, the people, the ruler', implying a relationship in which the people assume a central position so as to give legitimacy to the ruler and even to the Islamic project."[83] Until as late as 1990, Hermassi's writings remained generous towards a Tunisian Islamism that he viewed as being oriented toward modernity and pluralism: "A conceptual effort toward modernism, shaping the movement into an avant-garde, partly able to face the underground as well as the broad daylight of public life, and at last dictating the choice of a more convincing—and more and more convinced—tendency toward a legalistic strategy: this is the main feature of the Tunisian Islamicist movement." [84]

However, as the confrontation resurfaced between President Ben Ali's régime and the Islamists, all objective approaches to this issue became extremely difficult for Tunisian academics. Hermassi, for example, was appointed Tunisian ambassador to UNESCO, a position from which he could not continue to express publicly his views about *al-Nahḍa*, for they differed from those of the régime. Again, the state's considerations emerged as the stronger argument and the priority of most of the Tunisian élites.

In fact, the majority of politicians, academics and journalists have always shown suspicion towards the Islamists' liberal claims, if not total rejection of them. Much of their argument concerns the slogans used by the movement, such as the necessity of forming an Islamic government, or the implementation of *sharī'a* laws. Those who have been keen not to show a "negative attitude" towards religion have argued that putting Islam at the heart of politics is "not good" for Islam, because:

> It brings it down to the level of a temporary ideology, in an attempt to explain changing political choices, and because it holds it responsible for the mistakes of human actions. The right approach then is to keep religion as the closest uniting factor that brings all people of the Islamic nation together, regardless of their races, colours and political systems. It is unjust to religion to involve it in the social struggle to legitimise organised violence, instead of being a source of mercy and protection, and to [use it to] hide the faults of institutions ... [85]

Those who do not believe in the political role of religion from the outset have been more outspoken. They have argued that the religious state implies an absence of freedom of belief, "as was shown for centuries."[86] They have also pointed out that it implies the absence of political rights, because a religious government, as the self-proclaimed representative of Allah on earth, tends to persecute people. As for the Islamists' argument that they would form an Islamic government only after a popular mandate, their critics consider this undemocratic, because it would still force one interpretation of religion on the whole population, whereas there are many different religious groups in Tunisia.[87]

The alternative for a large proportion of the Tunisian élites is a secular state, because "politics is a social issue ruled by a human [i.e., not a religious] order."[88] As has been previously mentioned, this trend denies that the current Tunisian state is secular and argues for a new political order that will keep religion apart from politics. As one secularist writes: "The state in Tunisia has never been secular. The rulers of yesterday were behind the emergence of the obscurantist movement, and offered it all that it needed to grow. They made religion one of the main pillars of their ideology and used it for their political ends."[89]

With regard to the implementation of the *sharīʿa,* the overwhelming position taken by the centrist (being those closest to the government) and the leftist groups is one of emphatic rejection. Abd Allah ʿImāmi, who worked for the governmental Tunisian Information Bureau, ridiculed the slogans of the independent lists supported by *al-Nahḍa* in the 1989 general elections, especially those concerning the implementation of the *sharīʿa.* After pointing out that times had changed since the period of the Prophet and his Companions, when the punishment (*ḥadd*) for theft or the position against interest (*ribā*) was reasonable, he declared that there was no way that laws created for the past could be valid for the present: "This ignoring of the factor of time by insisting on tying the present to the past, and the refusal to observe the changes and complications that have taken place throughout the past years and centuries, are the main problems of *salafī* thought."[90]

For yet another critic, a writer for the official media, the *sharīʿa* contains little more than a limited set of rules, incapable of fulfilling the needs of modern society, arguing that in any case "its implementation is totally impossible because its conditions are impossible to meet."[91] Meanwhile, for the leftists, the Islamists' call to implement the *sharīʿa* has revealed "their terrible face," and confirmed them as representatives of the "dark side of our civilisation."[92]

These reactions reveal that most of Tunisia's influential élites are in agreement on the issues of the Islamic state and the implementation of the *sharīʿa.* They share a categorical rejection of the idea, even if it is the result of a democratic decision, which is why almost all of them have supported the government's restrictions on the formation of "religious" parties, and supported it even when the supposedly "civilian" *al-Nahḍa* party was refused legal recognition after dropping its religious title.

Although the Islamists' views have been in most cases vague and unspecific, the position taken towards them by the majority of influential élites, including those in power, in academic circles, in the parties of the left and in the media, has been taken a priori, on the basis of their principles and not their specific proposals or ideas. The modern élites of Tunisia are unrivalled, even in supposedly secular Turkey, in their radical and undemocratic stand towards the Islamists.

The Implications of the Islamists' Programme

Despite the extent of the radical opposition to the Islamists' main ideas, a close study of the exact implications of Ghannouchi's proposals for a Tunisian constitution reveals that the gap between the models of the republican and the Islamic state is not that great, at least theoretically. By Ghannouchi's standards, only two main modifications or additions are needed to make the 1990s constitution "Islamic": to insist that all laws of the state must be compatible with the laws of the *sharī'a*; and to widen the responsibilities of the existing official Islamic Council to include ensuring that all laws passed by the parliament or the president are acceptable in the eyes of the *sharī'a*.

Almost everything else in the constitution of the Tunisian state is in fact similar to what Ghannouchi has proposed in *al-Ḥurriyyāt al-'amma*. For example, Tunisia in the 1990s is under a presidential government with an elected parliament, which is exactly the system favoured by Ghannouchi. As for the laws governing the state, one can easily perceive in his works a number of very positive, democratic and liberal notions which, if applied properly, could make Tunisia a very democratic country indeed. Restrictions on the formation of political parties or the freedom of the press could be lifted, a demand voiced not only by the Islamists but also by the majority of politicians.

As far as religion is concerned, the government is currently responsible for its well-being, especially the building and financing of mosques and Islamic education, the latter being an official part of the national curriculum. An Islamist government might choose, in addition, to increase the budget for mosques or the time allocated to religious education; it might also insist that textbooks become more religious in their content.

The question of implementing Islamic *ḥudūd* may also be simpler than it appears from the rhetoric of competing political groups. *Ḥudūd* are applied, for example, in countries such as Saudi Arabia without causing any undue strain in Tunisian–Saudi relations, or even in Saudi–Western relations. Even in the Sudan, where Islamists are in power and the *ḥudūd* are applied, there have been very few cases in which any specific punishment has been executed, for the laws make it extremely difficult to apply the *ḥadd* unless a number of strict conditions are met. In fact, Tunisian Islamists are very pragmatic concerning this issue: in addition to stressing the very rare circumstances in which these punishments should be applied, they also point out that this is not a fundamental component of their programme. Ghannouchi explains it accordingly:

> To say that to carry out a range of *ḥudūd* conditional upon certain circumstances will hold the society back and cause its decline is a form of insult, and not a rational analysis. The main factors of revival are represented in certain fundamental values such as work, and in the mind which believes in reason, the dignity of the human being and social justice. These are the pillars of the essence of Islam, as is clear in its texts.[93]

One of the main areas of debate between Islamists and their opponents is the political and social rights granted to women in Tunisian laws. During Bourguiba's reign, Islamists were known to oppose the Personal Status Code because it banned polygamy and allowed adopted children to take the names of their new relatives, although they were prepared to concede these points in the hope that the government would grant them legal recognition. In 1988, Ghannouchi told a national newspaper that the Personal Status Code was "a body of choices and decisions which are part of different schools of Islamist thought."[94] This also reveals the pragmatic nature of the Islamist's thinking: a policy which was inherently unIslamic could become the product of "different schools of Islamist thought" if circumstances dictated.

A government led by the Islamists would also have to find an answer to the religious problem of producing and selling wine as part of the tourist industry, one of the country's main sources of hard currency. Wine is forbidden to Muslims, although the Islamists claim that they would still be flexible and keep it legally available for consumption by non-Muslims.

From the above comparisons, it is clear that the Islamist's Islamic state would not require the replacement of the existing legal system in its entirety. Changes in the constitution would be limited, and only the rhetoric and orientations of an Islamic government would be different, a fact that is rarely given mention in the ongoing war between the Tunisian government and the Islamists. It seems that their opponents have chosen to overlook this fact in an attempt to legitimise their own political interests.

The Question of Originality

If the Tunisian élites failed to recognise the intellectual efforts of Ghannouchi and his colleagues, could Ghannouchi possibly expect a more favourable judgment by foreign observers, either Muslim or non-Muslim?

One may safely say that a wide section of Arab Islamists in fact see Ghannouchi as the leader of the democratic trend within the Arab Islamic movement. In contrast to the conservatism of the Egyptian Muslim Brotherhood, Ghannouchi has been one of the most outspoken Islamists voicing the case for democracy. The issue of democracy in Tunisia has been high on his agenda since 1981, although it has not proved to be as important to other Islamic movements, who have faced different sets of challenges. This has given him the chance to be different, and to appeal to many Arab Islamists who view him as the voice of renewal within the Islamic trend in general. The editor of the magazine *Qirā'āt Sīyāsīyya* has described Ghannouchi as the representative of a "distinguished intellectual trend on the Islamic scene."[95]

Muḥammad Salīm al-'Awwā, one of the most distinguished contemporary Islamists specialising in Islamic government and author of the well-known work, *On the Political Order in the Islamic State*[96], wrote in the preface to Ghannouchi's *al-Ḥurriyyāt al-'amma* that "in all parts of the book, Rāshid has clearly sided with the Islamic position that supports freedom against dictatorship and progress against conservatism."[97]

This statement gives an indication of the essence of Ghannouchi's theoretical enterprise: it is a serious effort to collect the views of both past and contemporary Islamic scholars and writers on issues of freedom within the Islamic political system. In doing so, he proposed to

> choose the view with the strongest proof, the one that serves public interests and reflects the opinion of the majority of scholars, trying not to forget for a moment the current Islamic circumstances in which this message must be used, as a means of change and reform and of solving its problems. Our goal is not merely intellectual, but is an Islamic revolution that will uproot dictators from the land of Allah.[98]

There are, however, dangers in his methodology, in that it imposes a predetermined political objective on the research as a whole, ultimately making it a political, rather than an academic, undertaking. Ghannouchi himself has been frank about this from the outset. However, this may well have played a part in limiting the originality of his work, in that he has merely opted to select from previous recorded opinions concerning the Islamic state according to their suitability within the current political predicament.

In some parts of his work, Ghannouchi fails to hide his intense bitterness towards the current régime in Tunisia, which he describes as a coalition of "pseudo-modernist groups allied to the army."[99] In other instances he makes a number of unsubstantiated claims, such as his assertion that *al-Nahḍa* won the majority of seats in the 1989 general elections, without supplying any evidence to this effect other than his reference to "documents which will be revealed in the future."[100] (As we have already seen, official reports recorded that *al-Nahḍa* won around only 17 percent of the vote.) He also enthusiastically defends the Algerian Islamic Salvation Front, praising its victory in the aborted general elections of 1991 as "a chapter in the Islamic struggle which uncovers the true face of the secular élite in the Maghreb."[101] Indeed there are many examples in which he comes across more as a zealot addressing a public meeting, than an intellectual pursuing an academic undertaking.

Although writing a book to incite revolution is not in fact an academic goal, as Ghannouchi himself has admitted, nevertheless it remains an attractive enterprise for the thousands of Islamists who share Ghannouchi's dreams of an Islamic state. In one of the few reviews published about the book, there is praise for the fact that it comes as the result of a long period of contemplation in prison: "This is new proof that most of those timeless books by activist thinkers have been written either when entering periods of persecution (*miḥan*) or emerging from them."[102]

There are, however, two more serious problems implicit in this political approach to the subject of the Islamic state. The first is raised by Dr Ṭaha Jābir al-'Alwānī of the International Institute of Islamic Thought in the United States, who has commented on Ghannouchi's views on democracy and citizenship in Islamic society. He too argues that Ghannouchi is trying to appease Westernised secularists in the Islamic world, by claiming that Islam can accommodate a Western model of democracy, for political gain. Al-'Alwānī argues that this is little more than a continuation of the practice of imitation of Western civilisation, which has thus far been a complete failure. Thus, he asserts, "All thought which fails to make progress in the contexts of innovation and imagination, and in leading the nation from stagnation and traditionalism, will not make a relevant contribution. We may even say that it will fail, regardless of whatever short-term gains it may have."[103] Al-'Alwānī insists also that being faithful to true Islamic values should be more important than giving in to purely political considerations.[104]

This leads to the second problem in Ghannouchi's political interpretation of Islam, as it is presented in this book and indeed most of his other works: it is in fact politically motivated, and thus self-contradictory on more than one occasion. One of the more recent examples of this has been highlighted by Hermassi, concerning Ghannouchi's new approach in 1988 to the Personal Status Code, when defending the Islamists' political rights: "Having for a long time considered the Personal Status Code as part of a campaign of forced and alienating Westernisation, this same code is now considered to be a "body of choices and decisions which are part of different schools of Islamic thought." The code is an example of *ijtihād* (interpretation) and represents a positive element." [105]

It is interesting when one compares this with what Ghannouchi wrote in another of his books [The Muslim Woman in Tunisia], which was published in the same year: "The Bourguiban code known as the Personal Status Code was an anti-nationalist response to nationalist demands. It was not the result of Tunisian society's own evolution, nor a response to human demands and pressures. It was indeed part of a new crusader campaign to Westernise our society and to destroy its Arab-Islamic identity."[106]

Deciding which of the above two attitudes to believe concerning such a fundamentally important issue is problematic. One may, of course, adopt al-Zghal's approach, which argues that it is not uncommon for a politician to have different messages for different audiences, and that "the ambivalence of Ghannouchi's language is a reproduction of Bourguiba's language in the colonial era."[107]

Al-Zghal may have a valid point, although his approach is equally problematic as far as Ghannouchi is concerned, because it strips him of any religious legitimacy and sincerity. Even if the Tunisian Islamists are content to maintain their "political" face, this will not solve their problems of legitimacy, especially when one considers that they in fact lost the political battle with the Ben Ali régime in 1991.

Notes

1. John L. Esposito, *The Islamic Threat,* op. cit., pp. 163–4.
2. Al-Ghannūshī, *al-Ḥarakat al-ittijāh al-islāmī fī Tūnis,* vol. 3, op. cit., pp. 122–3.
3. Esposito, *The Islamic Threat,* op. cit., p. 159.
4. Al-Ghannūshī, *al-Ḥurriyyāt al-'āmma fī al-dawla al-islāmiyya,* p. 18.
5. See chapter 1, note 56.
6. Al-Ghannūshī, *Ḥarakat al-ittijāh al-islāmī fī Tūnis,* vol. 3, p. 84.
7. Al-Ghannūshī, *Ḥurriyyāt al-'āmma fī al-dawla al-islāmiyya,* p. 17.
8. Ibid., p. 17.
9. Ibid., p. 27.
10. Ibid., p. 19.
11. Ibid., pp. 86–7.
12. Ibid., p. 88.
13. Ibid.
14. Ibid.
15. Ibid., p. 98 and pp. 104–5.
16. Ibid., pp. 105–6.
17. Ibid., p. 99.
18. Al-Ghannūshī, *Maḥāwir islāmiyya,* op. cit. pp. 40–1.
19. A. Yusuf Ali, *The Holy Qur'an: Translation and Commentary* (Maryland: Amana Corp., 1983), p. 714.
20. Ibid., p. 24.
21. Al-Ghannūshī, *al-Ḥurriyyāt al-'āmma fī al-dawla al-islāmiyya,* p. 41.
22. Ibid.
23. Muḥammad Salīm al-'Awwā, *'An al-niẓām al-siyāsī fī al-dawla al-islāmiyya,* (Cairo: Dār al-Shurūq, 1989), pp. 210–16.
24. Al-Ghannūshī, *al-Ḥurriyyāt al-'āmma fī al-dawla al-islāmiyya,* op. cit., pp. 306–9.
25. Ibid., p. 44.
26. Ibid., pp. 44–68.
27. Ibid., p. 50.
28. Ibid., p. 109.
29. Ibid.

30. Ibid., p. 119.
31. Ibid., pp. 138–9.
32. Ibid., p. 139.
33. Ibid., p. 153.
34. Ibid., pp. 153–4.
35. Ibid., pp. 160–1.
36. Ibid., pp. 170–7.
37. Ibid., p. 247.
38. Ibid., p. 246.
39. Ibid., p. 247.
40. Ibid., pp. 247–8.
41. Ibid., p. 248.
42. Ibid., p. 342.
43. Ibid.
44. Ibid., p. 119.
45. Ibid., p. 247.
46. See *Ḥaqā'iq ḥawla ḥarakat al-ittijāh al-islāmī* (Tunis: n.p., 1983), pp. 10–12 and 16.
47. Ibid., p. 16.
48. Al-Ghannūshī, *al-Ḥurriyyāt al-'āmma fī al-dawla al-islāmiyya,* op. cit., pp. 298–301.
49. Ibid., p. 301.
50. Ibid., p. 304.
51. Ibid., p. 306.
52. Virginia: International Institute of Islamic Thought, 1993.
53. Rāshid al-Ghannūshī, *Ḥuqūq al-muwāṭana: ḥuqūq ghayr al-muslim fī al-mujtama' al-islāmī,* p. 72.
54. Ibid., pp. 66–8.
55. Al-Ghannūshī, *al-Ḥurriyyāt al-'āmma fī al-dawla al-islāmiyya,* pp. 292–4.
56. Ibid., p. 291.
57. Ibid., pp. 292–93.
58. Al-Ghannūshī, *Ḥuqūq al-muwāṭana* p. 77.
59. Ibid., p. 102.
60. Ibid., p. 112.
61. Rashid al-Ghannouchi, 'Participation in a non-Islamic Government', in *Power-Sharing Islam,* ed. Tamimi (London: Liberty Publications, 1993), p. 55.
62. Ibid., p. 54.
63. Ibid., pp. 54–5.
64. 'Abd al-Qādir al-Zghal in *al-Dīn fī al-mujtama' al-'arabī* (Beirut: Markaz dirāsāt al-waḥda al-'arabiyya, 1990), p. 340.
65. Ibid., p. 349.
66. Rāshid al-Ghannūshī, *Ilā 'ulamā' al-umma wa jamāhīrihā,* op. cit. See also *Al-Sha'b* newspaper, 18 September 1990.
67. Fahmī Huwaydī, "Ḥiwār ma'a al-Ghannūshī," *al-Majalla,* 3 October 1990.
68. Al-Darwīsh, *Ḥiwārāt,* op. cit., p. 112.
69. Ibid., p. 113.
70. See Ali Laridh's statement in *Réalités,* 26 October 1990.

71. See Morou's statement in *Réalités*, 17 May 1991.
72. See Morou's statement to *Réalités*, 12 April 1991.
73. Abdelkadir al-Jedidi in *Al-Ḥurria*, 9 June 1992.
74. See Abdelkadir al-Jedidi's open letter to *al-Nahḍa* leaders, *Réalités*, 10 July 1992.
75. Extracts from Ghannouchi's statement to the BBC, reproduced in *al-Nahḍa* publication *al-Ḥadath al-maghribī*, 23 May 1991.
76. 'Abd al-Qādir al-Zghal, *al-Dīn fī al-mujtama' al-'arabī*, op. cit., p. 344.
77. See Van Krieken's preface to *Khayr al-Din wa Tūnis: 1850–1881* (Tunis: Dar Sahnūn) 1988.
78. Khayr al-Dīn al-Tūnisī, *Muqaddimat aqwam al-masālik fī ma'rifat aḥwāl al-. mamālik* (Beirut: Dar al-Šalī'a, 1978), pp. 109–10.
79. Michael Brett, *Modernisation in Nineteenth Century North Africa*, p. 18.
80. Ibid., p.18.
81. Ṣāliḥ Bashīr, "Tūnis aw iḥtikār al-ḥadātha," in *Abwāb* magazine (Beirut: Summer 1994).
82. Zin al-Abidine Ben Ali, *Tunisia News*, 129, (Tunis: May 29 1995).
83. Al-Harmāsī, in *al-Ḥarakāt al-islāmiyya al-mu'āṣira fī al-waṭan al-'arabiyya*, op. cit., pp. 272–3.
84. Hermassi, "The Islamicist Movement and November 7," in Zartman, ed., *Tunisia: The Political Economy of Reform*, op. cit., p. 193.
85. Announcement by a group of Tunisian university teachers in *Al-Sabāḥ* newspaper, Tunis: 18 March 1988.
86. Al-Hammāmī, *Ḍidda al-ẓalāmiyya*, op. cit., p. 34.
87. Ibid., p. 34.
88. Announcement by a group of writers and artists in *Al-Mawqif* magazine, Tunis: 17 March 1988.
89. At-Tahir al-Hammami, "al-Lā'ikiūn yataḥarrakūna fi al-ḍaw"' in *Al-Sada* newspaper, Tunis: 3 April 1988.
90. 'Abd Allāh 'Imāmī, *Taẓīmāt al-irhāb fī al-'ālam al-islāmī: unmūdhaj al-Nahḍa: al-nash'a, al-taẓīr, al-haykala, al-irhāb ...*, op. cit., pp. 357–9.
91. Anas al-Shābbi, "Naqḍ al-mastūr wa ḥijābuhu 'an dawlat al-shara' wa du'ātuhu," *Al-Mustaqbal* newspaper, Tunis: 26 February 1988.
92. Al-Hammāmī, *Ḍidda al-ẓalāmiyya*, op. cit., p. 141.
93. Al-Ghannūshi, *Ḥarakat al-ittijāh al-islamī fi Tūnis*, vol. 3, pp. 128–9
94. Hermassi, in Zartman, ed., *Tunisia: The Political Economy of Reform*, op. cit., p. 199.
95. *Qirā'āt siyāsiyya*, 3–4, (Florida: Summer–Winter 1993), p. 121.
96. Cairo: Dar al-Shurūq, 1989.
97. See Preface to *al-Ḥurriyyāt al-'āmma fī al-dawla al-islāmiyya*, p. 13.
98. Al-Ghannūshi, *al-Ḥurriyyāt al-'āmma fī al-dawla al-islāmiyya*, p. 26–7.
99. Ibid., p. 274.
100. Ibid.
101. Ibid., p. 278.
102. Al-Fatih Abdessalam, "Civil Liberties in the Islamic State," in *Islamiyyāt al-Ma'rifa*, vol. 1, issue 2, (Malaysia: September 1995), p.173.

103. See Ṭaha Jābir al-'Alwāni's Preface to *Rights of Citizenship* by Rashid al-Ghannouchi, pp. 10–20.
104. Ibid., p. 11.
105. Hermassi, in Zartman, ed., *Tunisia: The Political Economy of Reform*, op. cit., p. 199.
106. Rashid al-Ghannouchi, *The Muslim Women of Tunisia*, (Kuwait: Dar al-Kalam, 1988), p. 148.
107. Abd al-Qādir al-Zghal, "The New Strategy of the MTI," in Zartman, ed., *Tunisia: The Political Economy of Reform*, op. cit., p. 216.

7

Issues of Identity and Westernisation

In spite of their deep absorption in politics, the Tunisian Islamists have always maintained that the issue of reforming Tunisia is primarily cultural. The first manifesto of the movement put as its two main goals the revival of an Islamic identity in Tunisia and the reforming of Islamic thought, in the light of the fundamental principles of Islam and the requirements of humankind's continual evolution and changing circumstances.

At the heart of this cultural question lies one of the most discussed topics in Tunisia and indeed in all the Islamic world: what Ghannouchi describes as "Westernisation, which has taken Tunisia from its roots, values and identity."[1] It may be argued that the challenge of the West has dominated political and cultural life in most Muslim countries since the French invasion of Egypt (1798–1801). As a consequence, political and cultural trends in the Muslim world have been divided into categories of "conservatives," "progressives" or "moderates," mainly based on people's attitudes towards Western domination and civilisation, and whether they chose to imitate, reject, or accommodate this within Islam.

As Hourani has observed, it was in as early as the 1860s that the issue of reform *vis-à-vis* Westernisation began to be clearly formulated by the élites of the Muslim world:

> It found expression in a movement of thought, directed in the first instance at the specific problems of the Near East, but raising once more by implication the general questions of political theory: what is the good society, the norm that should direct the work of reform? Can this norm be derived from the principles of Islamic law, or is it necessary to go to the teachings and practice of modern Europe? Is there in fact any contradiction between the two?[2]

One of the most prominent Muslims to formulate such questions was Khayr al-Dīn of Tunisia. Some of his views have been presented in the previous chapter with regard to the role of the state in implementing necessary reforms, and the need for Muslims to adopt those aspects of European systems and ideas deemed to be worthy according to Islamic principles.

Perhaps the most important observation which may be made about Khayr al-Dīn's approach is his insistence on the compatibility of those modern, Europe-inspired reforms with Islam. European progress, he argued, is not a Christian product, because Christianity does not interfere in politics and its prophet has advised his followers not to be associated with political leaders. If the Christian religion was the reason for this success then the papal state should have been in a much better shape. The real answer, according to Khayr al-Dīn, lay in "political institutions based on political justice, facilitating ways to create wealth, extracting treasures from the earth and benefiting from the sciences of agriculture and commerce. Most important of all these factors are security and justice."[3]

So is there anything anti-Islamic in this recipe for progress? On the contrary, argued Khayr al-Dīn, this is all part of the teachings of the *sharī'a*. *Shūra* is ordered in the Qur'an, and Muslims of the early ages led the way in decentralising power and encouraging the sciences and commerce, until their internal fighting led them into recession and stagnation. So, in a word, Khayr al-Dīn made it clear that "one of the main duties of Islam's rulers, ministers and scholars is to unite and build new institutions (*tazīmāt*) built up on justice and *shūra*."[4]

Khayr al-Dīn was unambiguous in insisting that the *sharī'a* was the main source of legislation for Muslims and that it remained valid for all times and places. In this respect, it is correct to describe him as an Islamic reformer:

> With regard to the *sharī'a*, Khayr al-Dīn remained, without a doubt, in the mainstream of classical Islamic thought. Since he saw the teachings of the *sharī'a* to be valid for all times and places, he then made it the basis for all of the *umma*'s activities, arguing that it was proof of the advantage that the Islamic nation had over other nations, for it contained rules and teachings which non-Muslims needed to learn and figure out gradually, sometimes paying a heavy price for that.[5]

The second characteristic of Khayr al-Dīn's thought was that it was pan-Islamic and not narrowly nationalistic, that is to say it was not confined within the Tunisian borders. At that time Tunisia was still a part of the Ottoman empire, although its rulers enjoyed a great deal of autonomy. Thus the *umma* in Khayr al-Dīn's terms meant the Islamic nation as a whole, and the Ottoman empire represented the political expression of this nation since it was the centre of the Islamic caliphate.[6]

In the light of the politics and culture of his time his rationale was not deemed strange, for most of the Islamic world was still viewed as one major political entity under the control of the Ottoman empire. In this sense, the modern notion of borders and national identity was still unobserved in the Muslim world.

Khayr al-Dīn served as prime minister in Tunis and later in Istanbul itself (1878–9). Eighty years later, however, by the time Bourguiba had become the first president of the first Tunisian republic, the Arab-Muslim world was totally different from that of the time of Khayr al-Dīn. Under the leadership of Atatürk, the Turks themselves had abolished the caliphate in 1924, ending all official forms of a united, pan-Islamic political entity. As a result, Turkey was converted to a secular state by its new constitution, guarded closely by the new republican army. Even its linguistic connection with the Arab world, the Arabic alphabet, was replaced with the Latin script by official decree. Atatürk argued that the caliphate had become both an irrational mission and a disaster, which the Turkish people could not conscientiously permit to continue.[7] In his view, the caliphate characterised the behaviour of a nation which looked ashamedly at its past, and was expressed in such minor details as the wearing of the traditional fez, as we may see from the following quotation:

> It was necessary to abolish the fez, which sat on our heads as a sign of ignorance, of fanaticism, of hatred to progress and civilisation, and to adopt in its place the hat, the customary head dress of the whole civilised world, thus showing among other things, that no difference existed in the manner of thought between the Turkish nation and the whole family of civilised mankind.[8]

Although at that time the notion of pan-Arabism seemed to present an attractive alternative to the abolished, once-united Islamic state, it was in fact the nationalist identity of separate and divided Arab countries that emerged victorious after the colonial period. For while a pan-Arab identity remained an aspired-to dream for many, the truth was that every Arab nation-state existed as a totally separate entity. Borders assumed a new significance as indicators of separate nationalities, and have remained significant—to the extent of being the main source of conflict among Arab countries since the 1950s.

In addition to the destruction of the united Islamic political entity, European domination of the Islamic world resulted in marginalising the reformist trend, which had attempted to introduce changes within the teachings of the *sharī'a*. This was to the benefit of a new class, which Sharābī has called the "Muslim secularists," being those of "the tendency to look forward, to progress, pragmatism, science, scientific doctrine, materialistic orientation of thought, dynamic views of social values, relativism of truth."[9] The Muslim secularists stood in contrast to the Islamic reformists who, despite being

> open to the values and achievements of modern science ... never really attempted to work out a systematic accommodation of it. Mostly they satisfied themselves with verbal compromises, tending always toward a defensive or apologetic stance. Instead of reformulating their assumptions in modern terms, they chose to follow traditional methods of rationalisation. Propaganda and polemics rather than criticism and analysis dominated their approach.[10]

In this post-caliphate Arab world, increasingly polarised between conflicting trends of thought and politics, Bourguiba was seen as a leading "Muslim secularist," as defined by Sharābī. In comparison with Khayr al-Dīn, Bourguiba was no longer attached to the state of the *sharī'a*, and was no longer committed to the pan-Islamic state. His attachment to the West, however, was more profound and strategic, shaping his views during the early stage of his political career, and before the independence of his country.

In 1951, he commented on the attitude of certain delegates at an Islamic conference held in India, at which he was present:

> There were those who, as a natural reaction to the West's mistakes and abuses, suggested the total and simple return to all that had made our great past, and, more precisely, to all that used to exist in our great past. They wanted the integral resurrection of the Muslim society of the first Caliphs, which in fact represented the golden age of Islam. It was an attractive and flattering suggestion, but one which is now impractical, even more so dangerous, because it would be translated as a fatal regression, throwing the Muslim world into a stage which had been golden in the past because it represented a formidable jump in the path of progress, but which would be nowadays largely outdated.[11]

A religious system of government was deemed to be out of the question for Bourguiba: "I have always thought that the Tunisian state, being non-theocratic and without a religious base, in that it includes non-Muslim citizens [Jews], can also assimilate Christians, who would be Tunisians in Tunisia without ceasing to be French in France."[12]

This attitude was also reflected in the manner in which Bourguiba handled the Islamic affairs of independent Tunisia, which, as we have seen in the first chapter, was one of the reasons behind the emergence of the Tunisian Islamic movement. What should also be added here is that Bourguiba was conscious not to be as brutal as Atatürk. In 1951, he wrote from Istanbul a short but telling remark: "I have seriously thought about this Kemalist experiment. There is [the question of] what to take from it and what to ignore. We can achieve the same results, or better, with less draconian measures that give more consideration to the soul of the people."[13]

His was a different form of secularism, in that it gave more emphasis to deeds rather than to words and slogans. None the less, without declaring an official secularist constitution, Bourguiba was indeed able to do much of what had been done by Atatürk in restricting the influence of religion in the state's affairs.

Another significant difference, which sets Bourguiba apart from Khayr al-Dīn, was that the former was a nationalist leader, with a vision of Tunisia as a political entity independent from the rest of the Arab and Islamic world. He saw Tunisia as a nation with its own independent history as represented by the state—which is an incarnation of the nation and which fulfils its hopes.[14] The battle for national independence from France also played a very influential role in giving legitimacy to this line of thought, for it was primarily a Tunisian issue, not Arab nor Islamic. Even when the Arab and Islamic dimensions of the struggle for independence were mentioned, they were used to serve the new nationalist concept in the face of French policies of assimilation:

> Where do we want to end? Is it serious to think that a nation of an ancient civilisation, Arab culture and Islamic religion; an ancient nation which had its hour of glory, a civilisation and a history; which had its own "state" ruled by a dynasty which itself came after four more ancient ones ... with the most ancient dating back to the eighth century, the era of Charlemagne and Harūn al-Rashīd; is it serious to imagine that such an individual nation, which is proud of its history, will allow itself to be absorbed and dissolved into a foreign community?[15]

However, this pride in an Arab-Islamic history, as a weapon with which to fight France, did not mean that "Tunisian-ness" should be diluted within a larger Arab entity. As Hisham Djait has observed: "Bourguiba's restrictive nationalism was strongly opposed to unifying Arab nationalism."[16] Indeed Bourguiba was also greatly influenced by the symbols of the pre-Islamic era in Tunisian history, especially those of Jugartha and Hannibal. More important than all of these experiments and symbols, however, was the fact that the Bourguiban era itself was proclaimed to be the stage at which Tunisia had finally expressed itself as an independent and sovereign entity, and as a safe and strong nation.[17]

A third significant dimension of Bourguiba's programme was his attachment to the West in general and to France in particular, a point which has been addressed in detail in the first chapter. Bourguiba was adamant even during the difficult period of the colonial era that the fight for independence was not against France *per se*: "In trying to put an end to the colonial régime, my actions have no other goal but the extermination of the germs of hatred between two brotherly peoples who, if this condition is met, will inevitably be involved in close co-operation, on the bases of solidarity and human fraternity."[18]

In fact, Bourguiba lived up to his ideas and promises. Under his leadership, independent Tunisia maintained a special and close relationship with France in the political, economic and cultural aspects. Many of his supporters agree that Tunisian francophony is "inseparable from its Arabism and Islamism. These three factors, along with three others, Mediterraneanism, Africanism and Maghrebism, make up the six basic reckonings of Tunisia's identity."[19]

In a general sense, Tunisia has always seemed to be more interested in the West than in her Arab neighbours, while the main goal that Bourguiba identified for his country was to catch up with the "civilised" nations, by which he meant those of the West. As a consequence, the Tunisian Islamic movement has challenged the Bourguiban régime on all three of the main issues mentioned herein: the relation between religion and politics (as discussed in the previous chapter), Tunisian identity and relations with the West.

Islam, Arabism, and Tunisian Identity

According to al-Najjar, who has no problem with the term "fundamentalism," the fundamentalist approach to the question of Tunisia's identity derives from the fundamentalists' interpretation of Islam in accordance with the model of the first Islamic state led by the Prophet Muḥammad. This implies that Islam comes first, in the sense of a wider Islam which covers and rules over everything.[20] Therefore, belief in Allah and the basic teachings of Islam is not merely a personal matter, as with other religions, but may only be fulfilled in its social context, whereby Islamic values become the basis for social relations and public life, an issue which is of prime importance when defining Tunisia's identity.[21]

Al-Najjar also asserts that the second defining factor of Tunisia's identity is that it be based on the moral system of Islam, which provides the Tunisian people with their own correct set of values. The Tunisian nation, he says, is still deeply committed to Islamic moral rules despite Western attacks, a fact which has enabled her to reject foreign customs.[22]

Third are the social and economic aspects of Tunisia's Islamic identity, which must also reflect those rules stipulated in the Qur'an, especially those of justice, equality, equal opportunities, and political freedoms. As an Islamic nation, Tunisia should obey the rule of Islam in all of these affairs, including family and criminal law.[23]

We may see from the above, therefore, that if Bourguiba argued that Tunisia could not have a religious government, the Islamists take precisely the opposite view. The government, says al-Najjar, is responsible for implementing the teachings embraced by the Islamic nation and adopting them as a basis for its public life. The nation should only chose its president to rule by Islamic teachings; if he fails to do so, then he must be deposed.[24]

According to al-Najjar, the final essential element in defining Tunisia's identity is Arabism (*'urūba*). He argues that Tunisia is essentially an Arab country which adopted Arabic as its language in the early stages of Islam, thus becoming an assimilated part of the Arab nation.[25] The main consequence of this, therefore, is that Tunisia must be proud of the Arabic language, using it for education and administration, and adopting Arab customs which are basically Islamic. It should be noted here, however, that this attitude does not suggest that the Islamists speak the language of Arab nationalists. In fact al-Najjar rejects "all forms of fanaticism" in defining the Arab identity, cautioning that one cannot boast of his race because the *sharī'a* forbids this, and has made it clear that only good deeds make a difference between people.[26]

In conclusion, al-Najjar argues that the Tunisian nation is a historical continuity of the Islamic nation, founded in that country in the first century after the *hijra*. Because of this, it possesses a theological and social heritage, being the comprehensive Islamic model transferred across the generations from the time of the Prophet. This heritage has remained alive in the nation's consciousness, despite all difficulties, which explains the popular appeal of Islamic renaissance movements.[27]

All of al-Najjar's ideas are embodied in the first major objective adopted by the Tunisian Islamic movement in its 1981 manifesto, which consisted of the revival of the Islamic identity of Tunisia so that it might resume its traditional role as a major centre of Islamic civilisation and culture. The last of the manifesto's five total objectives also touched on the same subject: it discussed contributing to the revival of the political and civilisational unity of Islam nationally, regionally (that is, within the Maghreb), within the Arab world and internationally.

As the Tunisian Islamists were not able to take the opportunity to govern or to be part of a government, their main battles remained on the level of principles, general aims and policies. This goes some way towards explaining why the identity issue was of such major importance in their campaign against Bourguiba's régime. Indeed, in some of Ghannouchi's statements, the identity question may be seen to have reached the level of personal revenge against a government that had shut its doors in the face of those educated in Arabic, yet supported a Westernised education system.[28] The contradiction he highlights is between two camps: those educated as Muslims and Arabs, and those favouring a Westernised education.

Ghannouchi argues that Arabism and Islam together are in opposition to the foreign culture and identity as propagated by Bourguiba. However, like al-Najjar, Ghannouchi's attachment to Arabism is only on the condition that it must not contradict Islam. In another statement he argues that the relationship between Islam and Arabism is very special and strong, as Islam is the founding element of the Arab nation. What is alien to *'urūba* however is secularism, and the Arab nationalist view which not only rejects Islam as the project of Arab unity, but also ignores Islam and even makes war against it.[29]

This is not a simple academic differentiation, but an issue of great importance in contemporary Islamic and Arabic thought and in the political debates of independent Tunisia. Arab nationalism as proclaimed in the Middle East was mainly secularist with a strong socialist inclination: such were the policies of Nasser in Egypt, and later the Ba'th parties in Iraq and Syria. The political mood of the sixties put socialism forward as a progressive ideology with which to fight against Western imperialism and Israel, and thus the Soviet Union became Nasser's main ally against the Western powers.

The secularisation of Arab nationalism was also seen as a means by which to unite all the Arabs of the Middle East, including sizeable Christian minorities in Egypt and Lebanon who, Arab nationalists argue, could not be mobilised under a strictly Islamic banner. In this context, *'urūba* is viewed as an alternative to the classic Islamic alliance which used to hold Muslim Arabs together with non-Muslim Arabs.

Generally speaking, however, a different approach has been taken in North Africa, for almost all the peoples of the region are followers of the Maliki school of Islamic jurisprudence. In this respect, Arab and Muslim have always been two close, non-conflicting, descriptions; indeed as we have seen, even during the short phase of his affiliation to Arab nationalism Ghannouchi claims to have "understood being an Arab and being a Muslim as inseparable realities."[30]

Like almost all major political ideologies, the secular concept of Arab nationalism moved from the Middle East to North Africa, where various small parties and groups of supporters became established; some supported Nasser while others supported the Ba'th ideology of Iraq. As with most other Tunisian political groups, the university campus was their main free space in which to express their views and to seek support from students. It was in this manner that the Islamists came to be embroiled in their bitter disputes with the Arab nationalists, despite their common cause against Bourguiba. The reality was that the two trends were competing for support among the students, and the nationalists generally found themselves closer to the secular leftist groups. The two trends seemed unable to find any common ground, not even over the Palestinian question, which ostensibly united all Arabs: the nationalists were known by their slogan *Falastīn 'Arabiyya*, in opposition to the Islamists' *Falastīn Islāmiyya*.

It is perhaps for this reason that the leaders of the Tunisian Islamic movement appeared to be somewhat cautious when, on presenting their manifesto in 1981, they were asked about their views regarding *'urūba:*

> If we mean *'urūba* as a matter of fact, meaning that we are Arabs and that Arabic is our language, then that is an objective reality. If some people mean *'urūba* as a social, political and economic philosophy, then they have to produce their versions and clarifications, because there are many conflicting views on that. We disagree with them, of course, and we do have our special understanding of *'urūba*.[31]

It was only after a decade that Ghannouchi began to show more enthusiasm for Arab nationalism, and he became an active preacher for a closer *rapprochement* between all Islamists and Arab nationalists, albeit still with the request that the nationalists abandon secularism. His argument was as follows:

> I have heard no Islamist object to the basic strategy of the idea of Arab nationalism. Every strategy is based upon a starting-point, and I believe that the starting-point for Arabist thinking is the unification of the Arab world. Political unity is the main idea behind nationalist thinking, and what came later was merely additions

influenced by international trends of thought, such as claiming that socialism is essential to *'urūba*—this supposedly essential element has helped to divide the Arab world instead of unifying it—or such as claiming that secularism is a basic element of *'urūba*, a claim that has deepened our divisions.[32]

Hence, whatever affiliation to *'urūba* the Tunisian Islamists may have shown, it should be understood only in the context of their battle against what they considered the secular and Western-orientated policies of Bourguiba. It was this approach that placed them in the same cultural camp as one of the most famous of Bourguiba's prime ministers, Mohammed Mzali, who in fact led the first political and security campaign against them in 1981.

The Role of Mzali

Although Mzali had little in common with the main ideas of the Tunisian Islamists, he was a well-known defender of the Arabic language and culture within the ruling Destour party, viewing them as fundamental components of the Tunisian identity. The most significant achievement of his long ministerial career was to Arabise most textbooks at all levels of education; a process he hailed as early as 1971 as a basis of Tunisian-ness.[33] In a speech he delivered to directors of primary-school teachers' institutions in the same year, Mzali insisted that they become more attached to their Tunisian roots and Arab-Islamic personality.[34]

Mzali was never an Islamic activist. He speaks excellent French and knows much about French literature from his studies at the Sorbonne. None the less, with his Arabisation policies he may turn out to have been the most important player in the history of independent Tunisia with regard to the country's Arab identity. He defended his approach even after becoming a political refugee in France in 1986, a country known to defend strongly the presence of French in its former colonies.

A studious pupil, I was aware of the benefits of the education I was receiving. For this reason, I did not reject it. But I continued to feel Arab and Muslim. I wanted to be a Tunisian and not second-class French. In a word, I wanted to be myself. My country existed before French colonisation, and will exist in the future, reinvigorated and back to its former self, when colonisation will have ended.[35]

This attitude was also reflected in Mzali's political career, as he describes in his own words:

As chief of the minister of national education's cabinet from April 1956 till May 1958 (under Mr Lamine Chebbi), and later as minister of national education myself, I was always a partisan of one clear option: Arabisation. Arabisation never

> meant xenophobia for me and my friends, or to retire within oneself, or a sort of return to the past. What I wanted—what we wanted—was that the Arab language would cease to be in exile within us, i.e., within itself, in an Arab and a Muslim country; and that it would restore itself to its normal rank: number one.
>
> We also wanted Arabic, adopted as the official language of independent Tunisia, to be capable of fulfilling its mission and to occupy the ground progressively in all domains of life: thought, the sciences and technology.[36]

Gradually, Mzali used his influence within the state to succeed in having most of his ideas implemented and goals achieved. By the time he was appointed prime minister in 1980, Arabic had come to the fore in the educational system in primary and secondary education, and its influence in the universities had augmented. Ironically, it was the emergence of the Islamists as a strong political force which complicated things for his policies of Arabisation. Some of those in the extreme secular groups around Bourguiba and the far left began to view Arabisation as a source of the Islamists' threat, referring to the close intellectual relationship between Arabic and Islam, for Arabic is the main key to the original texts of Islam. For some of his opponents, Mzali was even seen to be conspiring with the Islamists:

> The movement had its old relations with Mohammed Mzali. Articles *in al-Ma'rifa* gave him a lot of praise and applause, and he himself was very generous, working hard and being very dedicated during his time in the Ministry of Education to pointing out those educational programmes marked by poor ideas and mediocrity. He gave prominence to metaphysical thought and chased out every methodology or text that opened the mind and helped one to understand.[37]

Mzali was not one of Tunisia's most successful prime ministers, especially in running the economic affairs of the country and the oppressive manner in which he dealt with the trade unions. However, the main reason Bourguiba cited on sacking him in 1986 was that he had gone too far along the path of Arabisation.[38] It was a demonstration of the prominence that had been given to the Arab language, a helping factor in the cause of the Islamists. This dangerous connection become increasingly important for some as the political presence of the Islamists grew.

The argument was also used in a pan-Arab context, to discourage the use of Arabic in education; indeed Mohammed Arkoun argued passionately that critical thought becomes paralysed because of the holy connection between Arabic and the revelation. This, he explained, was why the Arab language had not participated in the debate about intellectual modernity, secularism, human rights and the study of religious phenomena. He also argued that this was why the political protests of the advancing Islamic movements exploited to the utmost all of the religious connections with Arabic.[39]

By putting the identity question at the top of their political agenda, the Islamists drew the attention of their opponents to the role of education in deciding the final outcome of the battle between their two different visions. This helps to explain why reforms for educational programmes were widely used in the campaign of Ben Ali's government against the Islamists. Mohammed al-Sharfi, an ex-communist selected to be minister of education, made it clear that education had been "used for many years as a propaganda tool for religious extremism and for the political manipulation of religion, and this is not the role of the school."[40] As a result, the national political charter adopted by the new régime became the deciding reference for religious education, and French regained some ground in the primary schools. The Islamists were frustrated by these reforms, which they described as being "hostile to the Arab-Islamic identity of the people" and as "enforcing Westernisation and secularisation."[41]

These verbal accusations were part of an all-out confrontation which soon took place between the régime and the Islamists, and which reinforced the polarisation between the two main approaches to the issue of Tunisia's identity. In one camp was the government and almost all the other political groups, opposed to all kinds of religious interference in public life, and worried by the Islamic movement's influence and ambitions. All of these parties stood together in defending the reforms in education and denouncing the protests of the Islamists as unjustified and similar to those Christian clerics responsible for the inquisitions of medieval times.[42]

In the other camp were the Islamists, who saw themselves, in the words of Ghannouchi, as "the true inheritors of their country's culture and glory." Similarly, they denounced those "who do not carry the message of Islam, as though strangers to this country," accusing them of being "the remains left by the colonialists after their withdrawal."[43]

Once again, we see how identity is used in opposition to the West and its "remains." It would be useful, therefore, to consider how the West is in fact perceived by the Tunisian Islamists.

The Fight Against Westernisation

The West has never been a secondary issue for the Tunisian Islamists in their fight to reinstate what they perceive to be their country's Islamic identity. In their view, President Bourguiba was leading Tunisia towards becoming a "Western" nation, a process to which they were determined to put a stop. Indeed, this may be considered the main target of the Islamists' struggle.

The question of Westernisation is of course not confined to Tunisia alone, but extends to almost the whole of the Muslim world. Even after most Islamic countries gained independence in the fifties and sixties, the legacy of Western colonialism remained extremely potent in the minds of leading politicians and thinkers, and was equally evident in the lifestyles of the people in the street. Elites were generally divided between those who wanted to imitate the West as a means by which to "move forward," and those who wanted to adhere to traditional and Islamic values, so as to resist Westernisation and bring about the rebirth of the "Islamic golden age." Those who were sensitive to the changes under way between the West and Islamic world chose to highlight this fact for a number of different reasons, although they all reinforced the notion of incompatibility.

Although there are those who have always maintained the possibility of mixing the good things of Islam with those of the West, the economic and political disputes between Western and Islamic countries have tended to give impetus to the more direct approach of portraying relations between the two blocs as a struggle between two civilisations. One Western proponent of such a view is Bernard Lewis, who argues that "when civilisations clash, there is one that prevails, and one that is shattered. Idealists and ideologues may talk glibly of 'a marriage of the best elements' from both sides, but the usual result of such an encounter is a cohabitation of the worst."[44]

From the Muslim side, most thinkers have acknowledged the impact of the clash between the Western and Islamic civilisations, although some of them—especially the Islamists—have insisted that their civilisation is not shattered and need never be, because the real problem lies with Western civilisation. Such was the argument of Sayyid Quṭb:

> The age of the white man's supremacy is over, because the white man's civilisation has accomplished its limited and short-term aims, and has nothing left to give to humanity in the beliefs, concepts, principles and values necessary to lead humanity, and to allow for the progress and prosperity of the human race, human values and human life.[45]

In Tunisia, Islamists have accepted that it is Muslims, not Islamic principles, that have failed during the last few centuries, and have insisted, like Quṭb, that Western civilisation has nothing substantive to offer the Islamic world for its progress and prosperity.

During the first few years of *da'wa* activity in the seventies, the Islamists criticised various aspects of Westernisation, especially education. As early as 1973, Ghannouchi expressed his fury over the philosophy textbook he was teaching at his secondary school:

> How can we present Sartre's theory on moral values and freedom as a general rule, without linking it to a specific historical period which the West went through following the destruction of its moral values, the weakness of its spiritual credentials and the disappearance of its supreme ideals of life—to the extent that he merely sees life as stupidity, anxiety and dullness. One has to ask: in whose interest are we destroying this generation, uprooting it from its cultural roots, cutting it off from its environment and leaving it lost, unable to define either the nation to

which it belongs, or the culture to which it relates, or the values by which to judge things? [46]

Ghannouchi's criticism here concerns the crisis of identity which independent Tunisia has undergone, and the uprooting of traditions by philosophers such as Sartre, and Western values in general. This is a problem that many others have also been able to discern, even from the side of Western analysts. Indeed, Bernard Lewis had articulated a number of complaints comparable to Ghannouchi's some ten years earlier, when analysing the impact of the West on the peoples of the Middle East. He recognised that, among other things, "the people of the Middle East ... lost their ancient corporate identity. Instead of being members of a millennial Islamic imperial polity, they found themselves citizens of a string of dependencies and then nation-states, most of them entities new to history, and only now beginning to strike roots in the consciousness and loyalties of their peoples."[47]

With regard to the social and cultural aspects, he also noted the following:

> The old patterns were destroyed, the old values derided and abandoned; in their place a new set of institutions, laws, and standards were imported from the West, which for long remained alien and irrelevant to the needs, feelings and aspirations of the Muslim peoples of the Middle East. It may well be that these changes were "necessary" and "inevitable," as these words are used by historians. The fact remains that they brought a period of formlessness and irresponsibility deeply damaging to Middle Eastern polity and society.[48]

While some sections of Muslim society have been willing to accommodate these changes, and even defend them as the only means by which to progress, the Islamists have remained unimpressed. They maintain that economic failures will not be addressed by merely adopting Western modes of living, but by asserting Islamic values. As Ghannouchi points out: "The economic problem cannot be solved other than by a humanistic approach to the issue. We must know: who *are* we? And to which culture do we belong?,"[49] thereby giving clear reference to the religious approach.

Islamists have answered Ghannouchi's above questions in two ways: directly, by asserting the Islamic nature of Tunisia's identity and culture, and indirectly, by discrediting Western values and culture, and insisting on its incompatibility with the values and culture of an Islamic country. This may in fact be the most prominent theme in Islamist literature since the early seventies. Although political considerations have determined the tone of the Islamists' negative attitude towards the West in the late eighties and in the nineties, the essence of this attitude remains unchanged, as is reflected by the way in which Ghannouchi has portrayed the West to his followers throughout the history of the Islamic movement.

In 1973 he wrote that the turning-point in the history of Western civilisation occurred when it came into contact with Islamic civilisation in Andalusia, and during the Crusades against the Islamic Middle East. Although the West was willing to learn from the scientific knowledge of the Muslims, it was also determined not to be influenced by their values, literature, language and concepts concerning the universe, life and the human being. Thus, Westerners insulted Muslims, their Prophet and their religion "to safeguard their civilisational personality from being dominated by Islamic culture."[50]

Four years later in *al-Ma'rifa* magazine, Ghannouchi devoted three long articles to a discussion of the same topic under the title, "Once again, We and the West." His starting-point in these articles was that increasing numbers of Islamists were calling for renewal and modernism at that time, an issue closely linked to the West as the main power, thereby making it imperative that a clear position on this point be given, so as to guarantee the correct direction for the Islamic movement to take.[51]

First, he rejected Albert Hourani's assertion that the Muslims' problems with the West were easing. Indeed, he asserted, the emerging Islamist movements were both an expression of their rejection of the West, and an indication of the awakening of Islamic civilisation as a means of solving the world's problems. He also pointed out that the industrial revolution, led by reason alone, had ended with the dragging of the "rebellious Western world against Allah to a failure more dangerous for the human being and civilisation than the previous one,"[52] by which he refers to the influence and the dominance of the church.

Quoting Socrates, Pavlov, Descartes, William James, Thomas Ben, Sartre, Marx, Kant, Nietzsche and others, Ghannouchi propounded a theory of the central idea of Western civilisation: "the belief in the human being as a master of the universe through his will, mind and technology."[53] While this philosophy has brought about positive results, freeing the Western man from his fears about nature, enabling him to rule over it and to explore new avenues for progress, to value freedom, to revolt against all dictatorships and to develop a democratic political system, he argues that its negative consequences have almost annulled the positive.

Ghannouchi also asserts in these articles that the West has misused scientific methodology by applying it to the human being and refusing to recognise its spiritual nature. As a result, science has failed to provide a meaning for the existence of humankind or its purpose in life, or a constant standard for its values. This has resulted in intense anxiety, listlessness, loneliness, confusion, and the feeling that life is frivolous, leaving the human being with the only option of seeking refuge "in dancing, drugs, or suicide to express his anger—or by joining one of the terrorist gangs."[54]

He goes on to argue that the supremacy of the principles of hedonism and utilitarianism have made society a theatre for the struggle for pleasure between the strong and the weak, between a minority which controls wealth and an exploited majority struggling to survive, thus destroying the principle of justice for the sake of freedom. This problem has expanded from Europe to the world, giving birth to colonialism, which also means that democracy has become a slogan by which to control the weak and to justify exploitation by the rich. In short: "the rebellion of the human being against his Creator in the name of freedom and independence has resulted in his giving away his soul and freedom."[55]

In an article written in 1979, addressed to local politicians fearful of the Islamic movement, Ghannouchi returns to the same subject. He attempts to posit the argument that the West is the only party with a real interest in fighting the Islamists, and that they in turn should not adopt this agenda. He argues that the West, driven by its heritage of a "crusade-complex" and its capitalist greed, with no concern for justice or freedom, is doing "all that is in its capacity to deepen elements of decadence in the Muslim world, to strip it of its identity, and dilute its values to thwart all chances of its renaissance and of regaining its glory, thus reinforcing its dependence on Europe."[56]

Part of this radical stance may be explained by the anti-Western mood created at that time by the success of the Iranian Revolution. Indeed, according to Ghannouchi, these same reasons were behind the hostile attitude shown by the West towards Khomeini's revolution in that same year.[57] What is more, when commenting on the views shared by those whom he regards as the three main leaders of the contemporary Islamic movement—al-Bannā, al-Mawdūdī and Khomeini (not including Quṭb or al-Turābī)—he mentions their liberation from Western culture, quoting Khomeini as saying:

> Culture is the basis of the happiness or unhappiness of the people. If culture is not suitable, then the youth educated within that framework will be misguided. Colonial culture produces a younger generation that can be colonised. It is more dangerous than the weapons of these tyrants. Our culture today is colonial, and it is not controlled by good people. [58]

In yet another article, Ghannouchi likens the West to a pharaoh who wishes to make all peoples of the world submit to his will, and, in the way that the biblical Pharaoh used magicians to try to discredit Moses, the West is using new charms—the media, interference with the news, facts and personalities—in its attempt to preserve its dominion over and exploitation of humanity's fortunes. In order to enjoy this status for as long as possible, the West also plans and works to "keep peoples in the position of the honest servant to his master, especially those in the Muslim world, who believe in a set of civilisational and human values which do not allow them to accept humiliation, defeat and dependency."[59]

In 1980, Ghannouchi formulated one of his main theses on the impact of Westernisation in an article entitled "Westernisation and the Determinism of

Dictatorship" (*al-Taghrīb wa Ḥatmiyat al-Dīktātūriya*). He began by reaffirming the failures of those ruling Muslim élites attracted to Western values and institutions in solving the social and economic problems of Muslim countries, claiming that this was because those alien values were in fact incompatible with the culture of the masses. Thus, it represented a failure which incited people to support all calls for change and revolution. However, in order to avoid being overthrown, and to safeguard their positions and interests, the Westernised élites became isolated from the masses, and found no alternative but to display the same disregard for their "Western culture, which considers freedom sacred, and to opt for means of repression—to confiscate civil liberties, to form repressive military and police services supported by other militia to protect themselves from the rebelling masses, to prevent them from exercising their rights in political organisations, and to deny them the freedom of press."[60]

According to Ghannouchi, this means that the policies of the Westernised élites have three main prongs: first, to confiscate civil liberties and to install a dictatorial régime; second, to form a new class that both benefits from the régime and is committed to it, living in small, closed societies in a totally Western manner, including the celebration of the West's religious festivals such as Christmas. These constitute small islands within the larger society, wherein the élites try to overcome their fears of the people's revolution by drinking and dancing. The third prong consists in ensuring strong and close ties with the West, receiving from it economic and military protection in return for protecting the Westernisation of their Muslim countries, and protecting the political and economic interests of the West in the region.[61]

By applying this theory to Tunisia, we may see that Bourguiba and his supporters represent the isolated Westernised élites, and the Islamists are the expression of the oppressed masses. As for the repeated references to revolution, it is easy to make a connection with the Iranian model, which Ghannouchi supported unreservedly. It is difficult not to recognise the internal logic of this theory, with many examples in the Muslim world standing in testimony to it. The real problem, however, is in accepting that Bourguiba and his party were, at the end of the eighties, an "isolated minority" in Tunisia, or in accepting that the Islamists were anywhere near to a majority at this time, or ready to lead a revolution. A more accurate analysis would be that Ghannouchi was comparing Bourguiba to the Shah of Iran, and the Destour party to the Iranian secret police or Savak, whereas Khomeini and his supporters where shown to represent himself and his followers.

This is a common problem with Arab Islamic movements, in that they often attempt to present subjective judgements in more objective terms. Another example of this could be the failure of the Egyptian Muslim Brotherhood to influence the standing of Nasser in the minds of most Egyptians and Arabs, even up to the present day. For them he represented little more than a dictator, an American agent, and even one of the enemies of Allah, who "accused the

Brotherhood of extremism, terrorism and manipulating religion to attain power; they fabricated events to legitimise the oppression of the Brotherhood, such as the plot of trying to assassinate Nasser in Alexandria."[62] For most Arabs, however, Nasser was and still is seen as a nationalist leader and a combatant against Western imperialism.

Equally, although Bourguiba was always Western-oriented, he was primarily a Tunisian leader, and was seen by many as the most successful Tunisian politician of the twentieth century. His legitimacy was not built on Western protection, but, more importantly, on his leadership of the national independence movement and his party, which was active in almost every part of the country even before independence. This may even go some way towards explaining the drastic manner in which he dealt with certain religious issues, such as his call for workers to stop fasting in Ramaḍān. At that time he was not counting on Western protection, but on his popularity as the leader of the nation, and on his claim to act as a Muslim leader exercising *ijtihād*.

The problem with Ghannouchi's assessment, therefore, is that it ignored this connection—however weak—between the government and what he called "the culture of the masses." Other analysts have been able to identify this point, as we may see from the following quotation:

> The governments that were formed after liberation were mostly secular, and even the Moroccan King, Muḥammad V, promoted an image of himself as both a religious leader and head of a modern state. Algeria and Tunisia used very pronounced Western models for their political systems. All these regimes professed Islam as a reformist frame of reference rather than as the basis of a political order, and this position was widely accepted.[63]

According to this logic, the problem for the ruling élites, therefore, was not only Westernisation, but also "the fact that they were all firmly entrenched and have remained in power for more than a generation, [which] eventually led to some forms of dissent, and the use of Islamic terminology to express it, an imitation of a growing trend throughout much of the Islamic world."[64]

Ghannouchi's logic was different, however, and was articulated once more on behalf of the whole Tunisian Islamic movement in the 1981 manifesto. Its main goal was the rebirth of the Islamic identity in Tunisia, so as to combat the prevailing dependency, alienation and disorientation. Another goal was to purge the effects of Westernisation from Islamic thought. One suggested method by which to bring this about was to "liberate the Islamic conscience from civilisational defeat by the West."[65]

However, it is remarkable to note that in the 1988 manifesto for the new *al-Nahḍa* party, the movement removed every reference to the words "West" or "Western" from the entire document, despite its being much longer and more detailed. Instead, the manifesto contained a more realistic political rhetoric concerning the operation of a balanced foreign policy based on mutual respect with other nations.[66] It is evident that direct references to the West had been dropped here for political considerations, for the leadership was trying to present a more moderate case to both the government and Western observers in Tunisia. Later, Ghannouchi also softened his attitude; the most significant expression of this mood came after going into exile in France, and later in Britain. In this instance, Ghannouchi accused those who viewed the West as merely an imperialistic power in the following terms:

> The West is not only that. The West is liberating thought, the experience of a liberating revolution, scientific progress, improvement to health and to the human being's information concerning the universe and people. In the West too there were true, progressive forces that stood by the Algerian revolution and stand currently with the Palestinian revolution. There are also in the West human organisations from whom oppressed Muslims can only find support. The West too has traditions of freedom, to the extent that oppressed Muslims who cannot find refuge in the vast Muslim countries seek asylum in Western countries.[67]

With Ghannouchi still to apply for political asylum in Britain, the personal dimension of this account was obvious. He went on to suggest that a future war between Islamic and Western civilisation was not necessary because, among other things, both were based on "common religious roots, the belief in the unitarianism of Allah, the hereafter and the unity of the Prophets."[68] He also suggested, "the two closest civilisations in history are that of Islam and that of the West."[69]

Speaking to a number of American Muslims in the United States in 1989, Ghannouchi seemed no longer to believe in the total bankruptcy of Western civilisation, as the following quotation illustrates:

> We are asking you, as Muslims living in a society that has reached the highest scientific level, not to imitate and reproduce the religious practices nurtured in underdeveloped environments. We ask you to be a new contribution to the Muslim notion, to reform Muslim thought, and to produce an evolved form of religious practice that will favour the evolution of underdeveloped Muslim populations. You are living in a country closer to Islam than underdeveloped countries, because here a set of norms, divine in the universe and in society, is well respected.[70]

This was a change of emphasis—to say the least—in Ghannouchi's views *vis-à-vis* the West. His negative rhetoric had given way to a more positive and conciliatory approach, with political aims in mind: to secure the safety of Islamists seeking asylum in France and Britain in particular, and to gain a certain degree of support for the movement in the looming confrontation with Ben Ali's régime. With the possibility of seizing power stronger than ever before, Islamists might have also found it necessary to start allaying the fears of Western politicians, and to persuade them to listen to what they would hopefully deem to be a modern and moderate Islamic party.

The problem with this change was that it was again inconsistent, and thus bound to be seen as a superficial stand that was undertaken only for tactical political considerations. Just two years after his visit to the United States, Ghannouchi was to speak his mind about the West once more—not in the way in which he had been doing since the seventies, but rather in a manner never displayed before. The occasion was also unusual, for it was February 1991, when the United States and its allies were preparing for the final assault to drive Iraq out of Kuwait. Ghannouchi, who then sided with what he called "the brotherly people of Iraq, its strong army and courageous leadership,"[71] opted to appeal to the basic religious sentiments of Arabs and Muslims. He denounced those who rejected the faith as "the camp of *kufr*," a term he had rarely used publicly before in discussing the West or indeed any non-Muslim nations. His theory was as follows:

> The world is going through a phase in which conflicts have moved from being inside the camp of *kufr*, after the collapse of the socialist bloc, to a phase of battling against Islam and its *umma*. The army of *kufr* assemble under the flag of America and Zionism, and shoot us with the same arrow, to enforce their total domination over our fortunes, and to put an end to our revival and hopes for progress. The camp of *kufr*, led by the United States, was angered by Iraq purely because of the latter's decision to go beyond the limits forced on Arabs and Muslims in the fields of power, industry and independence of will and decisions. Our acceptance of the Western plan, which uses the slender cause of liberating Kuwait to destroy an Islamic country that has tried to rebel against Western lordship, would be tantamount to a clear-cut recognition of the *kāfirīn*'s sovereignty.[72]

By using the religious terms *kufr* and *kāfirīn*, used in the Qur'an to define those who reject the Islamic faith, Ghannouchi was appealing to the religious sensibilities of the Muslim masses, and trying to lend a religious dimension to the second Gulf War. Gone here were Ghannouchi's political or intellectual interpretations of the West as either ally or imperial aggressor, or a progressive or disoriented civilisation; rather we see the words of a politician aiming to play the religious card by the most powerful means possible, perhaps in the knowledge that many Muslims, and non-Muslims too, were convinced that the war was primarily about oil. Oil was the main motive behind Iraq's invasion of Kuwait, and it was the same motive for the West in liberating Kuwait.

While Ghannouchi may be seen to be too politically driven—to the point of contradicting himself—al-Najjar has always been more academic in his approach and faithful to the main ideas of the Islamists concerning the West, albeit initially and essentially formulated by Ghannouchi himself. Al-Najjar has observed that leftist and secularist approaches to the issue of Tunisian identity have aimed at producing a materialistic society, interested purely in fulfilling the material needs of its members, and ignoring the religious question as a matter for individuals. This vision, he claims, has been borrowed from a model of philosophy and civilisation that is foreign to the cultural system of the Arab-Islamic nation, that is to say the Western model, whether of the Western or Eastern blocs.[73]

According to al-Najjar, the problem with Western civilisation is that it is based on one dimension only—that is, materialism—whereas the human being also possesses a spiritual dimension. This, he declares, fundamentally contradicts the nature of the human being, and with disastrous consequences, for it preaches death as the end of existence: "This colours life with gloom, as hope retreats to the moment of death, resulting in a bitterness which the human being feels whenever he finds the chance to think amidst his daily engagements."[74]

Like Ghannouchi before him, al-Najjar insists that life in the West is dull, hopeless, driving people to drugs or suicide, and implies that there are no goals left in life according to Western philosophy. This, he affirms, being the model which Westerners have adopted and to which they have become accustomed, it is now being forced on the Tunisian nation, despite being alien and far removed from its nature.[75]

Al-Najjar is unwilling to acknowledge any positive achievements of what he labels "the representatives of Westernisation" in Tunisia or elsewhere in the Muslim world, referring to those secular élites who took power on independence, only to imitate the West and serve its interests. The rule for him is that secularism is tantamount to a total political and social failure, and that one of the strongest examples of this is Tunisia:

> The Bourguiban régime adopted a secular identity in ruling the country, from which he formulated all of his programmes for politics, culture, education and the economy, making it the only ideological reference for every programme and action taken by the state, in a flagrant and unjust manner. After a third of a century of putting the Bourguiban project into practice, the results came close to an all-out disaster: an economy that had gone through many setbacks and ended in bankruptcy; an education system from which generations graduated, altered in their identity, their personality torn between an appeal rooted in their hearts in the heritage of the Islamic-Arab identity, and that which had been forced on them from Western culture; a society which forced hundreds of thousands of its unemployed youth beyond its borders to fall prey to all kinds of international crime, to the point that Tunisian youth are now included on the lists of criminals in all European countries; and social fragmentation, which has increased over the years and resulted in civil war, resolved only moments before explosion.[76]

Two main points emerge from the Islamists' views on Westernisation and its effect on Tunisian identity: first is their negative stereotyping of Western civilisation as hopeless and driving people to despair, drugs and suicide; second is their extremely negative evaluation of the work of the Bourguiban régime, as an "enemy" of what they view as the historic identity of Tunisia. There are, of course, what may be described as more moderate or balanced assessments of the West in some of the political statements issued by *al-Nahḍa* during the mid-1990s, or in some of Ghannouchi's articles and speeches, especially those he has written since he became an exiled politician resident in London. These assessments are few, however, and are not sustained, often seeming too politically motivated. What is more, as mentioned earlier, many are easily contradicted by extreme rhetoric, depending on the audience.

Assessments such as those put forward by al-Najjar and Ghannouchi may seem somewhat oversimplified, notably those of the West and Bourguiba. For example, for many Westerners and indeed also for many Muslims, the idea that Western philosophy is leading Westerners to hopelessness is exaggerated, and reflects an inadequate understanding of Western societies. It appears to be a view based on the accounts of court hearings published in the media, although these do not differ greatly from those news stories published concerning cases in Arab and Islamic courts. Further, although secularism has taken root in the West, one should not underestimate the continuing importance of religion in society. People still attend church, and many more who are not churchgoers still believe in God.

Indeed, the issue of secularism in particular is much more nuanced than Ghannouchi and al-Najjar have suggested, and many Western nations are quite ambivalent or ambiguous about the role played by religion in their societies. For example, some nations have established churches or state-sponsored religious schools, whereas in others, such as in France, where the state is apparently more overtly secularist in its ideology, there is a higher rate of religious observance than exists in Britain. In short, there are so many different indicators to be taken into account, and the issue is not as clear-cut as it appears in the Islamists' literature.

This does not imply, however, that the West does not face serious social problems—especially the instability of the family and the widespread use of drugs among the youth. These problems, however, like numerous others, are not confined to the West alone, for in many Arab countries, drug abuse is as serious as in the West.

There are two factors which may go some way towards explaining the Tunisian Islamists' categorical rejection of the West. The first is religious, in that the Islamists view Tunisia's real identity as being purely Islamic, regarding Western influences as a serious threat. The second is connected to the Islamists' religious and political fight against Bourguiba: presenting Bourguiba as a Western agent was seen to be an effective method of discrediting him as a Muslim ruler. Therefore everything that was a product of Bourguiba's rule was rejected by the Islamists, even if Islamic in nature, as we see below:

> The struggle of the Islamic movement, rather than being against paganism in the individual, is against paganism in the society. Their most visible opponent has been the Tunisian (Bourguiban) regime, as the force held responsible for the de-Islamization of the Tunisian society. The Islam of the Tunisian government is referred to as official Islam, as opposed to the *jihād* of the Islamic movement. The former, which is taught in the schools and communicated through newspaper, radio, and television applies only to the individual's religious life. For example it concerns prayer, fasting, ritual washing, family relationships and laws of inheritance. Struggling (militant) Islam, on the other hand, is based on the concept of comprehensiveness, a rejection of the separation of religion and politics in Islam and an insistence that Islam presents a program for all areas of the society's life. Struggling Islam, therefore, is in direct conflict with the regime, which is represented as the cause of various societal problems through its neglect of true Islam.[77]

Signing the National Pact

This extreme hatred of Bourguiba, as a symbol of secularism and Westernisation, may also explain why the Islamists later accepted and signed what amounted to a "collective response" to the question of Tunisia's identity under the supervision of President Ben Ali in 1988—two years before they were pitched in battle against him. This response was embodied in the National Pact, which was signed by all major Tunisian political parties, including a representative of *al-Nahḍa*, on the first anniversary of the 1987 overthrow of the leadership. The first item in the document addressed Tunisia's identity, providing a formula which appeared to reconcile all parties, even though it appeared—certainly from the rhetoric mentioned in this chapter—that this was an impossibility.

None the less, all parties agreed from the outset that the identity of the Tunisian people "is specifically Arab and Islamic, rooted in a glorious, remote past, which aspires to face up to the challenges this epoch offers."[78] This in turn was immediately linked to other aspects of Tunisia's identity, for the document asserted: "The fact that our country is situated in a region which was the cradle of many great human civilisations, has enabled our people to contribute to human civilisation over the centuries and has fitted it for renovation and creativity."[79]

The text of the National Pact was a reflection of the difference in emphasis between the Islamists' insistence on the priority of the Islamic dimension of Tunisian identity, those insisting on Arabism, and those leftists and members of the ruling party who were unwilling to give in to the arguments of the Arab nationalists and, more particularly, the Islamists. It is also on account of these differences that the document elaborated on its main themes by referring to Tunisia's pre-Islamic history, when

> Carthage was one of the two greatest powers of the ancient world. Our people are proud of Hannibal's genius and Jugurtha's heroism. Tunisia is proud of having been

> the starting point for conquests which brought the message of Arab Islamic civilisation to the Arab Maghreb, to the North of the Mediterranean and to Africa. Tunisia is also proud of the geniuses it has produced, such as Imām Saḥnūn, Ibn Khaldūn the scholar and Khayr al-Dīn the reformer.[80]

In the same delicate way in which it balanced Jugurtha with Imām Saḥnūn, the document went on to explain the main implications in considering Tunisia's Arab and Islamic identity: Arabism was portrayed as being prior to Islamism, despite the fact that the first was a consequence of the second. Further, it was carefully explained that there would be no attempt to undermine foreign languages, seen by secularists and leftists as the gate to the modern sciences, thus declaring:

> The national community is required to strengthen the Arabic language, by making it the language of communication, administration and teaching. Of course, we must be open to other civilisations and languages, particularly those used in science and technology, although obviously the national culture can only evolve in and through the national language. Here we must avoid any split between the élite and the masses, because this may well emasculate the élite, and isolate the masses from modern life.[81]

The discourse of this text revealed that the views of the secularists were stronger than those of both the Arab nationalists and the Islamists, in that emphasis was not given to Arabic as a historical language of the sciences. On the contrary, the need for foreign languages was mentioned specifically as being a tool for the study of science and technology, meaning that the Arabic language could not, and could never, play such a role. Interestingly, "foreign languages" continue to be used for sciences in Tunisia.

For many who compare this with other nations, who are content to use their national languages both to teach and learn the sciences, the argument against Arabic seems absurd, and to be motivated essentially by an inferiority complex *vis-à-vis* the West. Furthermore, the use of "foreign languages" in the plural form here is misleading, because the language referred to is French alone and no other. The problem with this lies in the fact that the use of French was forced by purely political considerations, linked to France's historical influence over Tunisia. If the issue rested merely on the study of the sciences, or a means by which to communicate with the rest of the world, English would have been a far more useful language for Tunisians to use. Pre-empting this line of criticism, the document also promised: "We must work to develop the national language to make it a language of science and technology, able to handle contemporary thought, whether innovative or creative, and to contribute—and rightly so—to human civilisation."[82] For the nationalists and Islamists, however, the inevitable problem with Arabic was that it had no chance of becoming the language of the sciences in Tunisia unless it was used as such with full commitment by the state.

With regard to Islam, the Islamists were compelled to add their names to a document that did not preach the comprehensiveness of religion or the rule of the *sharī'a*, and which specified that looking after Islam was the role of the state:

> The Tunisian state will watch over the noble values of Islam, as the touchstones by which Islam may constitute a source of inspiration and pride, and be open to the concerns of mankind, the problems of this day and age and modern life, and for Tunisia to remain what it has always been—one of the centres of Islamic influence and a focus of science and *ijtihād*, thus perpetuating the avant-garde role once played by Kairouan and by the Zeitouna University.[83]

The final part of the document constituted the most blatant point scored by the government and its allies against the Islamists concerning the issue of Tunisia's identity. In the light of "the deviation from Islam" which the Islamists had been used to seeing under Bourguiba, they were persuaded—or were forced—to accept one of the most significant examples of Islamic *ijtihād*, only one year after the changes of November 1987. As a result, the Islamists added their names to a document which asserted the following:

> The Code of Personal Status and the laws that complete it came after independence, introducing a package of reforms. The most important were the abolition of polygamy, the granting to women the right to marry without a guardian once the age of reason had been reached, and the ushering in of equality between men and women as regards divorce and its procedures. These reforms aimed at freeing and emancipating women, in line with the age-old hope that existed in Tunisia, based on the solid rule of *ijtihād* and on the aims of the *sharī'a*; they prove that Islam is vital and open to the requirements of the modern ages and of evolution.[84]

As if to assure Western observers and the opponents of the Islamists that the state had not made any kind of concessions to the Islamists, the document then concluded by stressing, "the Tunisian state strengthens this rational orientation, which springs from *ijtihād*, and works to help *ijtihād* and reason, having a clear impact on teaching, religious institutions and the means of information."[85]

An overall assessment of the National Pact's response to the question of identity shows that it was more interested in laying down safeguards against the Islamists' influence than in explaining how the Islamic identity of the country should be expressed. Further, it did not make any apology to the Islamists for Bourguiba's policies, and although the former president was not mentioned by name in the document, his legacy was represented in the text as being the embodiment of progressive Islamic *ijtihād*.

Despite all this, the Islamists accepted the pact and signed it. However, it is obvious that their acceptance of the document stems from overriding political considerations. At that time, they were concerned mostly with giving whatever guarantees were necessary to the government in order to be recognised as a political party, even if this meant making concessions on matters that were at the very heart of their campaign against Bourguiba. They may also have argued that they were merely putting their signatures to a document that could be amended later, a move which could also prevent the restriction of their political activities, if they later managed to gain legal recognition.

The signing by the Islamists of the National Pact also marked the decline in the value of charters and commitments in Tunisian political life. It was to prove that the Islamists were not alone in their way of thinking when, only three years later, the text of the National Pact covering democracy, human rights, freedom of the press and the multi-party system had been virtually ignored by almost all of Tunisia's political parties in the midst of the confrontation between the government and *al-Nahḍa*.

All that now remains are hollow words, void of any serious commitment, responsibility or obligations, and what seems to be in store is merely a bitter and continual war between the state and the Islamists over, among many other important issues, the unresolved question of Tunisia's true identity.

Notes

1. Al-Ghannūshī, *Ḥarakat al-ittijāh al-islāmī fi Tūnis*, vol. 3, op. cit., pp. 122–3.
2. Hourani, *Arabic Thought in the Liberal Age*, op. cit., p. 67.
3. Khayr al-Dīn al-Tūnisī, *Muqaddimat aqwam al-masālik fi ma'rifat aḥwāl al-mamālik*, op. cit., p.116.
4. Ibid., p.160.
5. Van Krieken, *Khayr al-Din wa Tūnis*, op. cit., p. 131.
6. Al-Tūnisī, *Muqaddimat aqwam al-masālik fi ma'rifat aḥwāl al-mamālik*, op. cit., p. 110.
7. Mustafa Kemal, *A Speech* (Leipzig: K.F.Koehler, 1929), pp. 592–3.
8. Ibid., pp. 721–2.
9. Hisham Sharabi, *Arab Intellectuals and the West: The Formative Years 1875–1914* (Baltimore: Johns Hopkins University Press, 1970), p. 10.
10. Ibid.
11. Habib Bourguiba, *La Tunisie et la France* (Paris: Julliard, 1954), p. 267.
12. Ibid., p. 331.
13. Ibid., p. 289.
14. Ali al-Ganari, *Bourguiba: le combattant suprême*, op. cit., p. 293.
15. Bourguiba, *La Tunisie et la France*, op. cit., p. 342.
16. Hishem Djait, *La personalité et le devenir Arabo-Islamiques* (Paris: Seuil, 1974), p. 26.
17. Wannās, *al-Dawla wa al-mas'ala al-thaqāfiyya fi Tūnis*, op. cit., p. 125.
18. Bourguiba, *La Tunisie et la France*, op. cit., p. 287.

19. Baccar Touzani, *Maghreb et Francophony* (Paris: Economica, 1988), p. 78.
20. Al-Najjār, *Ṣrā' al-huwiyya* fī Tūnis, op. cit., p. 33.
21. Ibid., p. 34.
22. Ibid., p. 36.
23. Ibid., p. 37.
24. Ibid., pp. 40–1.
25. Ibid., p. 41.
26. Ibid., p. 42.
27. Ibid., p. 74.
28. Al-Ghannūshī, interview, *Arabia*, April 1985.
29. Al-Darwish, *Yaḥduthu fī Tūnis*, op. cit., p. 59.
30. Al-Ghannūshī in *Arabia*, op. cit.
31. Al-Ḥāmidī, *Ashwāq al-ḥurriyya*, op. cit., p. 206.
32. Rāshid al-Ghannūshī, *al-Ḥiwār al-qawmī al-dīnī* (Beirut: Markaz dirāsāt al-waḥda al-'arabiyya, 1989), p. 341.
33. Muḥammad Mazālī, *al-Fikr*, no. 16, 7 April 1971.
34. Muḥammad Mazālī, *Dirāsāt* (Tunis: Tunisian Company for Distribution, 1984), p. 120.
35. Mohammed Mzali, *Tunisie: Quel avenir?*, op. cit., p. 139.
36. Ibid.
37. Al-'Imāmī, *al-Nahḍa*, op. cit., p.189.
38. Mzali, *Tunisie: Quel avenir?*, op. cit., p. 11.
39. Muḥammad Arkūn, *al-Fikr al-islāmī: naqd wa ijtihād* (London: Dār al-Sāqī, 1990) p. 17.
40. Mohammed al-Sharfi, at a press conference reported by Le Maghreb, 13 October 1989, p. 18.
41. From an *al-Nahḍa* communiqué, published in *Le Maghreb*, op. cit., p. 9.
42. Al-Hammāmī, *Ḍidda al-ẓalāmiyya*, op. cit., p. 12.
43. Al-Ghannūshī, *Maḥāwir islāmiyya*, op. cit., p. 36.
44. Bernard Lewis, *The Middle East and the West* (London: Weidenfeld and Nicolson, 1963), p. 43.
45. Sayyid Quṭb, *al-Mustaqbal li hadha al-dīn* (Kuwait: IIFSO, 1978), p. 56.
46. Al-Ghannūshī, *Maqālāt*, op. cit. p. 12.
47. Lewis, *The Political Language of Islam*, op. cit., p. 44.
48. Ibid., pp. 44–5.
49. Al-Ghannūshī, *Min al-fikr al-islāmī fi Tūnis*, op. cit., p. 22.
50. Al-Ghannūshī, *Maqālāt*, op. cit., pp. 23–24.
51. Ibid., p. 41.
52. Ibid., pp. 43–4.
53. Ibid., p. 51.
54. Ibid., pp. 55–6.
55. Ibid., p. 56.
56. Ibid., p. 66.
57. Ibid., p. 80.
58. Ibid., p. 93.
59. Ibid., p. 110–11
60. Ibid., p. 168–9.

61. Ibid.
62. Muṣṭafā Mashhūr, *Tasā'ulāt 'ala al-tārīkh*, (n.p., 1983), p. 68.
63. Alan R. Taylor, *The Islamic Question in Middle East Politics* (Boulder: Westview Press, 1988), p. 84.
64. Ibid.
65. Al-Ghannūshī, op. cit., pp. 7–8.
66. This document has been published in al-Ghannūshī's book *Ḥurriyyāt al-muwāṭana fi-l-waṭan al-islāmī*, op. cit., pp. 339–48.
67. Darwīsh, op. cit., p. 151.
68. Ibid., pp. 152–3.
69. Ibid., p. 154.
70. Abdelkadir al-Zghal, "The New Strategy of the Movement of the Islamic Way," in Zartman, ed., *Tunisia: The Political Economy of Reform*, op. cit., p. 209.
71. Rāshid al-Ghannūshī in the unpublished communiqué "This is the Day on which the Truthful will Profit from their Truthfulness."
72. Ibid.
73. Al-Najjār, *Ṣirā' al-ḥuriyya fī Tūnis*, op. cit., p. 105.
74. Ibid., p. 106.
75. Ibid.
76. Ibid.
77. Magnuson in Zartman, ed., *Tunisia: The Political Economy of Reform*, op. cit., p. 178.
78. The National Pact (Tunis: Tunisian External Communication Agency, 1992), p. 9.
79. Ibid.
80. Ibid.
81. Ibid., p. 10.
82. Ibid.
83. Ibid.
84. Ibid., p. 11.
85. Ibid.

Conclusion

On attempting a retrospective analysis of the history and discourse of *al-Nahḍa*, one finds oneself necessarily addressing two fundamental questions, both of which at first sight may have been deemed irrelevant. The first question is "What is *Islam*?," and the second is "What is *an Islamic movement*?"

Many would maintain that attempting a definition of Islam should not be problematic. None the less, it has proved to be a major source of contention for those movements which have dedicated their efforts to defending Islam and implementing it. That is because they are talking about Islam as a political remedy for the crises of the Islamic world. Indeed, as one former member of *al-Nahḍa* has noted, "you will grow tired of looking for a clear and comprehensive answer to this question within the abundant literature of the Islamists ... generally speaking, Islamic discourse today has no clear outlines even to its defenders, let alone to the élites and to the public."[1]

With reference to the Tunisian Islamists, possibly the most prudent way in which to address the above question is to concede that almost all aspects of the "Islam" dealt with in this thesis have had political associations. Whether under the banner of the call for an identity independent of the West, or for the "enforcement of liberties" and the overthrowing of the Bourguiban régime, Islam has always been presented within a theoretical framework for political solutions, and as a mode for articulating anger and demanding change.

The intellectual project of al-Afghānī and ʿAbduh at the beginning of the twentieth century seemed irrelevant to the new activists of Islam. It was people such as al-Bannā, al-Mawdūdī, Quṭb and Khomeini who, at a later stage, offered the guiding models for the Tunisian Islamist. Thus they favoured the activist approach to Islam, rather than the reformist and cultural approach of ʿAbduh, or even the pan-Islamism of al-Afghānī, based on mobilising the whole Islamic community rather than dividing it into "good" and "bad" Muslims. If ʿAbduh, the grand Imām of al-Azhar meant little to the Tunisian Islamists, it is to be expected that Khayr al-Dīn al-Tūnisī, even more inclined to the West, should receive very little attention from the activists in the Tunisian Islamic movement.

It is also interesting to note that, in spite of all of its claims, the Tunisian Islamic movement has rarely concerned itself with the classical model of Islam, as represented in classical Islamic literature. It may also be argued that, even if the movement had genuinely wished to have reference to the classical Islamic model, it would have been incapable of doing so, because there was no single reputable scholar among the movement's leadership. Although it is fair to say that during the formative years of the movement there were hints of an exclusively religious philosophy, since 1979, however, the movement as a whole has pursued a one-way route to the heart of politics.

According to the Islamists, politics is an integral component of Islam, thus there is nothing wrong in placing politics high on their agenda. Hence we see that even when dealing with the very rudiments of Islam, they are interpreted politically. In an attempt to summarise his movement's understanding of Islam, Ghannouchi states that Islam embodies a comprehensive methodology for liberation: "it liberates humanity from the tyranny of dictatorship and exploitation; it is a call to unitarianism and its attendant values of equality, fraternity, freedom and the love of justice."[2]

It should none the less be observed that Islam does not exclusively concern itself with politics. Indeed, it may be argued that politics is a "grey area" in Islam, and that specific political forms have been left broadly to Muslims to define for themselves within a set of general values, as Ghannouchi himself admitted in his attempt to define the model of an Islamic state. For the sake of political expediency, however, *al-Nahḍa* chose not to concentrate on the fact that Islam is also a human appeal to humankind and a source of social legislation. Indeed, since applying for legal permission to operate as a political party, *al-Nahḍa*'s non-political concerns of morality, faith and social harmony have almost totally disappeared from their agenda, and the aims of the organisation are geared largely to taking political power.

It appears that this problem may not be confined only to the Tunisian Islamic movement; rather, it is a reflection of a wider and a more general mood among the majority of contemporary political Islamic movements, who tend to be inclined towards a primarily political mission. In Islamic terms, this has been an unbalanced approach, as many observers have noted:

> Though they enjoyed greater political support, the resurgence organisations had a unidimensional approach to Islam that in most cases had little humanistic content and emphasized the importance of political solutions. Attempts to develop hybrid doctrines combining neofundamentalist principles and a number of other concepts of proper order proved to be chimerical in terms of implementation. The situation that finally emerged as Islamic resurgence and which became part of the political equation in the Middle East was one in which ideology became more an instrument in the struggle for power than a blueprint for genuine change in the future.[3]

This leads us to our second question on the definition of an Islamic movement. Ghannouchi's response is that its main characteristics are "its comprehensive understanding of Islam, and its active approach to building an Islamic society based both on that comprehensive understanding and the establishment of an organisation working towards the building of an Islamic state."[4] His explanation asserts that the belief in a comprehensive Islam is the essence of an Islamic movement, otherwise an "active approach" and the establishment of an organisation are little more than the human choices of a group of individuals.

There are a number of problems implicit in Ghannouchi's definition. Above all, what he describes as a "comprehensive understanding" of Islam is dismissed by many as being narrowly political; second, what he describes as an "active organisation" is really a political party that may, in the opinion of other Muslims, sometimes bring more harm than good to the Islamic cause. What is more, if one puts their claims to Islamic identity aside, there is nothing particularly religious about the organisational features and tactics of *al-Nahḍa.*

For instance, one could cite several examples of political opportunism. There was the occasion when, probably for internal reasons to keep both its secret and public leaders satisfied, *al-Nahḍa* declared that Ghannouchi was still its president, whereas this was not true. There was also the fact that it presented itself to the public as a peaceful civilian movement, meanwhile building a secret military wing which made two unsuccessful attempts to topple the Tunisian régime. There was also the instance, during Ben Ali's era, when they opposed the changes Bourguiba had made to the Personal Status Code, later only to sign their names to these changes, accepting them as Islamic *ijtihād.* One may also mention their support for Saddam Hussein as leader of a legitimate *jihād* against the West, and their subsequent retreat and attempts to appease the Gulf states. Finally, one could also mention the period when rival factions of the movement published opposing publications in Paris, each claiming to represent the party and discrediting the other; and when Morou described Ghannouchi as a man of violence and a *persona non grata* in his party. It is difficult to identify any specifically religious dimension to any of these adopted actions and positions.

Apparently the movement's attitude is that for as long as it continues to present itself as Islamic, then it is to be regarded as Islamic, irrespective of its deeds. This is perhaps more flagrantly obvious in other countries such as Afghanistan and Algeria however, where Islamic movements are implicated in killings, assassinations, and the bombarding of cities, yet are still viewed and identified as "Islamists."

Further, the main claim of the Tunisian Islamists to a comprehensive Islam and the dedication to building an Islamic state was founded on one very important assumption, which is the absence of an Islamic state among Muslim countries, at least until the Iranian Revolution. It is from this that other specific concepts have been derived, such as that of *jāhiliyya* in the society and the difference between the Muslim and the Islamist, claims that are very disputable in the majority of Muslim societies, as one observer notes:

> If the *sharī'a* was about a specific political system, then it would have disappeared with Islam itself at the end of the Ottoman era. The heart of the matter concerns society and its source of legitimacy, and the fact that Islam still occupies that place. If we claim otherwise, then that would mean that Islam has ceased to exist, despite the hundreds of millions of those fasting and praying, and the dozens of millions of pilgrims and worshippers, and despite the declarations of hundreds of millions of people that they believe in Islam.[5]

It was partly as a consequence of such objections as those expressed above, along with other political considerations, that *al-Nahḍa* dropped its "Islamic" tag in 1988, and that it was compelled to stress on a number of occasions that it did not claim to be the official voice of Islam. However, if the Muslim society is already in existence, then the Islamists' insistence on an Islamic state is rather along the lines of a partisan political programme that may appeal to some Muslims but not to all of them. This may imply also that when some Islamic movements speak of an Islamic state, they are in fact speaking of their aim to come to power, and do not mean that Islam is the monopoly of their members and supporters.

This in itself begs the question of whether one may really speak of an "Islamic movement," or a political organisation that claims to be Islamic within an Islamic or Muslim society. Indeed, this is a very delicate issue and concerns the relationship between religion and politics in Islam in general. None the less, despite conflicting statements about the rejection of a monopoly of Islam, neither *al-Nahḍa* nor most other Islamic movements have been able to resist the temptation to present themselves as the true defenders of Islam, so as to gain sympathy and support from the public.

It was stated in the foregoing chapters that the history of the Tunisian Islamic movement was to a large extent the fruit of its founder's experiments, trials and errors. It was also influenced by the ideas of *al-Tablīgh* movement, the Muslim Brotherhood, the Iranian Revolution and even extreme leftist groups in the university. The Islamic allegiance was so wide and vague as to accommodate the non-political preachers of *al-Tablīgh* and even the unbelievers among young Tunisian communists. Even major turning-points in the movement's history appear to have been somewhat precarious and unplanned. Foremost here is the fact that the announcement of the formation of *Ḥarakat al-Ittijāh al-Islāmī* in 1981 was in response to the arrest of Salah Karkar and Ben 'Issa al-Dimni in December 1980.

This in turn may emphasise the need to re-examine the "Islamic" appellation of the Tunisian Islamic movement, because it places Islam at the mercy of the successes and failures of a secret political organisation. This does not seek, however, to undermine the importance of *al-Nahḍa* as a political group that has succeeded in raising a number of issues of particular concern to a sizeable section of Tunisian society. Indeed, the movement's success was illustrated in the

1989 elections, when, according to official statistics, the independent candidates supported by *al-Nahḍa* gained up to 17.5 percent of the total vote, a figure that was probably higher according to the movement's leaders.

In brief, it seems that there is more than one reasonable argument to suggest that the Islamic appellation does not in fact reflect the true nature of the Tunisian Islamic movement as a political group attached to the notion of an Islamic state. In essence, it operates as a political group, and as such people will inevitably differ about the religious nature of the movement's actions. For example, is it acceptable, from a religious point of view, to do something in secret and announce something different to the public? Some may call this simply lies, but others may argue that it is a permissible manœuvre necessary for political considerations. As a purely political party attached to the principle of the Islamic state, however, *al-Nahḍa* should be judged by its deeds, successes and failures.

Mohamed Abid al-Jabri's remarks concerning Islamic groups in Morocco may apply equally to most other Islamic movements, including *al-Nahḍa*:

> Islamic groups that aim to bring about the "Islamic state" in Morocco, meaning that they want to acquire political power in it, will only succeed in establishing themselves among the masses as a force able to influence history when they adopt political and social goals which reflect the yearnings of the people and their spiritual and material needs. If a group does so, it will definitely change to a political movement, and thus its success will depend on its ability to achieve compatibility between religion and politics, meaning its ability to innovate in the field of religion according to the needs of the time, in such a way as to produce a religious discourse of contemporary social and political content.[6]

To a great extent, *al-Nahḍa* aimed to achieve the above, by listening to the needs of the youth and the trade unions, by opposing what it saw as the Westernisation of society and the political system, and, most importantly, by giving priority to the implementation of a liberal democracy in their declared political agenda. Thus there is evidence that many of the movement's slogans effectively expressed the mood of those large numbers of Tunisians who had become disenchanted with the *status quo*. The Islamists' successes may also be viewed within a wider Arab trend in support of the Islamists' cause:

> The Islamic resurgence is a response to the confusion and anxiety of modernity and a challenge to repressive and corrupt regimes. Like Christians during the Reformation, the Islamists attempt to reach directly the literal world of God and provide legitimacy to popular demands to transform their societies. Indeed the political clout the Islamists now have is due not to the desire of Arabs and others to live under strict Islamic rule, but to the perceived failure of Western models of political and economic order, including nationalism and socialism, to solve the Middle East's problems.[7]

It should also be noted, however, that making a sound academic connection between the Islamic reference and the grievances and aspirations of the people has proved to be very difficult, not only for *al-Nahḍa* but also for most contemporary Islamist groups. With the exception of Ghannouchi's work on the issue of the Islamic state, *al-Nahḍa*'s literature on innovation and *ijtihād* is very poor indeed. During more than twenty years of existence, the movement failed to make any substantive contribution towards a theoretical formula for one of its main goals: the reformulation of Islamic thought, taking into account the fundamental principles of Islam and the requirements of humankind's continual evolution and changing circumstances, as mentioned in the 1981 manifesto.

In reality, it is inconceivable that the movement could have offered a work of this nature, when it cannot boast one major scholar among its ranks. Not one of the leaders of the movement may be described as an academic expert on Islamic affairs, with the exception of al-Najjar, who was always a peripheral figure and was never regarded as one of *al-Nahḍa*'s leaders, to the point where he withdrew his membership in 1995. Another obvious reason for the movement's failure in this area is the increasing prominence of political issues on the leadership's agenda since 1981.

An obvious question here is whether the movement has compensated for its failures on the intellectual side by succeeding in its prime concern: politics. One possible response is that the movement has in fact risen to become the most influential opposition party in Tunisia, and that, in one way or another, it was instrumental in bringing about Bourguiba's downfall. Before becoming embroiled in their confrontation with Ben Ali's régime, this role played by the Islamists was hailed by Ghannouchi:

> The movement made friends with those who defended it, so it managed to isolate Bourguiba even within the secular circles in Tunisia, to the point that no party, syndicate or association hesitated to side with the Islamic movement. The movement also isolated Bourguiba on a large scale outside the country. This bloody struggle ended with the fall of Bourguiba, and that was not at at all easy for a man who had led a nation within a national movement, had led it in the battle for independence and in the building of the state. [In spite of this] ... the day of his fall became a day of celebration. "*Lā ilaha illā Allah; Bourguiba 'adū Allah*" [there is no god but Allah; Bourguiba is the enemy of Allah] was the most important slogan heard on the streets of Tunis during those seven months, in a bloody confrontation between the Islamists and the régime. [8]

Not all observers would accept, however, that the Islamic movement was the main factor in bringing about the end of the Bourguiban era. It would be fairer to agree with Hermassi, who asserts that *al-Nahḍa* was "the first opposition movement to stand up to Bourguiba," and that "there were many Tunisians during that fatal summer who began to feel that something had changed, and that

Tunisia would never be the same again."[9] Indeed, one should not forget that it was in fact Ben Ali who evicted Bourguiba and succeeded him to power, bringing an end not only to the confrontation between Bourguiba and the Islamists, but also to a succession problem that had lasted for ten years. One should not forget either that the 1987 political changes were faithful to the main Bourguiban values; indeed, the new régime showed a more concerted and vigorous attempt to improve Tunisia's relations with the West, and to combat all sources that could have been of help to the Islamists' cause.

In another important note, it is now clear that the political endeavour of the Islamists did not result in substantial change to the one-party system, or indeed to one strong-man rule in Tunisia. Lars Rudebeck argued in the 1960s about the Parti Socialiste Destourien as a "mass party" that either participates actively in or supervises the application of political decisions, and one that is "more effectively becoming the motor of the state while Habib Bourguiba remains its chief conductor."[10] That was a reflection of a situation when some of the top ministers and politicians were able to argue with Bourguiba and discuss issues with him.[11] But things changed in a radical way during the late 70s and in the 80s when the cult of the one undisputed leader reached the point of making Bourguiba the only real source of power and direction for the whole political system. What used to be the "mass party" became a mere component of the state bureaucracy, nothing more. That is why it needed a strong man from inside to take the colossal risk of evicting Bourguiba from office. And when Ben Ali succeeded in his gamble the party was left totally out of the equation: he could have destroyed it by a decree without facing the slightest hint of opposition from anyone. But he opted to continue using it as part of his political instruments to control political life after changing its name.

Ben Ali inherited a situation where absolute political power resided with the president in the elegant palace of Carthage. His confrontation with the Islamists placed him in even stronger need of tightening up the president's control over the entire political domain in the country. With the Islamists crushed, other political parties were unable to influence the decision making process in any shape or form. A few of their members made it to the parliament for the first time, but still within the limits of defending the president's policies. When the president and secretary-general of the MDS showed signs of dissent they were dealt with swiftly: arrested, jailed and their party divided into two factions. The parliament itself, the renamed ruling party and other independent associations were all made into "servants of the state," a description used by Clement Henry Moore to describe the relation between the then PSD and other special-interest organisations with the state during the first four years of independence.[12] Making the additional clarification that the "state" meant the president, at least throughout the second half of Bourguiba's long term in office, is not a great surprise. Again, it was obvious from the early years of independence that Bourguiba had been "a sovereign who rules as well as reigns. He has quasi-absolute power as head of state and as head of the party.... As the undisputed ruler, Bourguiba has retained and developed his role as the educator of the society.... As the ruler of a Muslim country, he claims the right to reinterpret Islam."[13]

This was the political system Ben Ali inherited from Bourguiba, and it is clear that the president's role was only revitalised by the energy of a much younger and strong-minded leader.

One of the ironies of the last major battle of *al-Nahḍa* was the fact that it took the name of "enforcing liberties," while the result was to re-emphasise the state's orientation to deal more firmly with dissent and opposition. It does seem indisputable that the political climate was more open and tolerant during the first two years of Ben Ali's reign, and that the sharp turn towards more authoritarian rule arose in the midst of the government's campaign to end, in a decisive way, the challenge and threat of the Islamists.

The Islamists have disputed the accusation that they played a direct or indirect role in pushing the new regime to be less tolerant of opposition and unwilling to share power with other parties. They point to the fact that secular parties which supported the government's campaign against *al-Nahḍa* ended up being attacked themselves, with some of their leaders joining Islamist prisoners in gaol, and concluded that the new regime of Ben Ali came with an agenda to suppress opposition parties regardless of what *al-Nahḍa* did or did not do during the years 1988 to 1991.

One thing is certain about this issue, and that is that one-party rule in Tunisia, if not the one strong-man rule established during the Bourguiba era, has re-emerged with vigour and strength, unaffected in any serious way by the political contribution of the Tunisian Islamic movement.

Many now think that Ben Ali's crackdown on the Islamists has driven them from the Tunisian political arena for a long time to come. For the government at least, the problem of what it calls "religious extremism" has been solved for good: repression against all members and supporters of *al-Nahḍa* is fierce, and even for those who flee overseas, the government does not spare any effort in having them extradited. Nevertheless, there are still reasons to be cautious in dealing with such claims, and these are to do with the shortcomings of repression everywhere. Hermassi described Bourguiba's policy against the Islamists as a failure, because, as he put it "repression was worse than a crime, it was a mistake (in Tallyrand's formula); it did more to reinforce than to weaken the Islamicist organization."[14] For many Islamists at least the rule is still valid, and Ben Ali's repression may be counterproductive in the long term.

It is widely accepted that Ben Ali's régime has dismantled the whole Islamist organisation inside Tunisia, and has criminalised any kind of support for their cause. In spite of this, no one can claim with any certainty that support for the movement has been uprooted from the hearts of its former supporters, and it is still conceivable that the Islamists could function as a strong political party, should the Tunisian government permit them to operate legally.

Predictions aside, however, the experiments of the Tunisian Islamists have given further reason to call into question the strategies of many of those other Arab-Islamic movements which have opted to employ double-speak and to pursue a double agenda, and which count on gaining power with the support of the military. For despite all the precautions taken, the Tunisian government's intelligence services consistently succeeded in discovering *al-Nahḍa*'s secret organisations, their hidden strategies and, most important of all, their military wing. In such circumstances, such a movement should recognise that infiltrating the army is a high-risk strategy in any part of the Arab world, with the most common punishments on discovery being the death penalty and the total dismantling of the movement.

Long before any initial confrontation with the Bourguiban régime had occurred, the Tunisian Islamic movement had set up the first cell of its military wing, imitating the actions of the Egyptian Muslim Brotherhood before it. On the one hand, it sought members for a reformist civilian movement and presented itself to society as such, whereas on the other it was preparing for a potential military coup. It was only from police and court records that many leading Islamists discovered that they were members of an organisation that actually possessed a military wing. Some may declare that this was dishonest behaviour on the part of the organisation towards its members and towards society as a whole, whereas others would see it as the only feasible method with which to topple the régime and bring about an Islamic state.

Unfortunately for the Tunisian Islamist movement, as for the Muslim Brotherhood previously, the military wing brought little more than disastrous consequences. While in Tunisia it provided the régime with an excuse to disband forcibly the entire operation, in Egypt this wing was also responsible for "placing the organisation [i.e., the Muslim Brotherhood] in a number of political crises, for providing the excuse for the dismantling of the movement, and for being hunted down and chased out because of the irresponsible acts in which it had become implicated."[15]

One other important point that should be made pertaining to the experiments of *al-Nahḍa*, is the great risk taken by the movement in having only one real leader speaking out on all issues and dealing with all of the movement's affairs. Throughout the entirety of the movement's history, Ghannouchi managed to remain its indisputable leader, even when he was not its *de facto* president. With Ghannouchi as the sole authority on almost all issues, the movement shared some of his limited successes, but also suffered from his unsuccessful political manœuvring and found itself finally in an impasse when, after he left Tunisia in 1989, Ghannouchi declared war against the Tunisian régime and sided with Saddam Hussein in the Gulf War.

As neither a professional politician nor a professional religious scholar, Ghannouchi constantly experimented with new political tactics according to Tunisia's changing circumstances. What is more, in making the movement and the leadership one and the same thing, we may see that in the nineties the two continue to be unified, and indivisibly so. Once again, the Tunisian Islamists were not alone in this, for most of their Arab colleagues placed their destinies in the hands of one undisputed leader who controlled the movement until his death. Perhaps ironically, such behaviour is imitative of that of current Arab rulers, who leave office only as the result of death or military coup.

Because the movement was so closely tied to Ghannouchi, its intellectual contribution came largely from him alone. An evaluation of his contribution is open to different interpretations: most of those in other Arab Islamic movements as well as some Western observers appear to acknowledge his keen interest in democracy, identifying it as one subject that is absent from other Islamist movements' discourse. For others, however, Ghannouchi's Islamisation of Western democracy does not stand alone as an original contribution to contemporary Islamic thought, for it has been addressed before by other scholars, although generally with only a limited degree of success.

In all cases, there remains a major question arising from this study: was the politicisation of Islam beneficial to the cause of Islam in Tunisia? Did it make people more attached to its teachings and more observant of its orders? Did it increase its influence as the main frame of reference for the whole population? Although it is difficult to give clear yes or no answers, one can still maintain that using religious motives and slogans to reach power, even by planning twice for a military coup, may lead to making Islam itself a subject of polarisation and conflicting arguments between political players within the society. From being the unifying religion of almost the whole population, which was the role of Islam during the struggle for independence, the politicisation of religion may eventually divide the nation along religious lines, on the assumption that good Islamists were fighting bad or secularist Muslims. The reaction of the Islamists' opponents in Tunisia offers an indication of what could happen as a result of this approach. Religious behaviour may become a cause of suspicion, and going to *fajr* prayers at the mosque might be interpreted to mean that the person concerned is a member of an "illegal" opposition party.

The damage caused by this political and cultural bi-polarisation could have been contained by a strong presence of the traditional class of scholars, the *'ulama'*. This was not possible, however, since the public role of the Tunisian *'ulama'* had started to fade a long time ago, as early as the start of French colonial rule in 1881. The French used various methods to neutralise Muslim scholars. However, it was President Bourguiba who undertook the dismantling of any potential independent base for them by closing down the *shari'a* courts and Zeitouna University and by abolishing all religious endowments (*awqāf*). In his book, *The Tunisian Ulama 1873–1915*,[16] Arnold H. Green presented a thorough work on the attack faced by the Tunisian *'ulama'* from the French colonial authorities and from the rise of young nationalists more inclined towards Western secularism. However, a general remark he made about secular-oriented Muslim governments' attitudes towards the *'ulama'* is reasonably valid in the Tunisian case:

> When governments confiscated endowments and abolished tax-farming concessions, many religious leaders lost the major sources of their wealth; when ministries were reorganised, the *'ulama'* were at once bureaucratised and excluded as political advisors; when Western-inspired law codes and courts were introduced, the *'ulama's* judicial powers were suppressed or severely reduced; and when secular schools were established, the *'ulama's* monopoly over education was broken and there developed a new Western-educated elite which tended to replace the old religious elite as the people's spokesmen.[17]

Tunisia currently has no equivalent to Al-Azhar university in Egypt or the dedicated Islamic universities in Saudi Arabia. The radical secular elements of the Tunisian elites may see this as a historical gain that shall never be reversed. But others may see it as the basic factor which legitimises the emergence of an Islamic movement, not only the one which seems crushed and dismantled by President Ben Ali, but also future ones which will always be tempted to cover the ground left by the *'ulama'* and settle historical scores with the secularists in government.

Will these threats of deepening political and cultural divisions convince the Islamists or their opponents to change their tactics? This seems very unlikely, unless Tunisia and other similar Muslim societies find their own peaceful way of handling political and social differences, and making political changes at the top possible without monopolising religion or staging military coups.

Notes

1. Ṣalāḥ al-Dīn al-Jūrshī, "Mustaqbaluhā rahīn al-taghyīrāt al-ḥadhariyya," in al-Ḥaraka al-islāmiyya: ru'ya mustaqbaliyya, ed. 'Abd Allāh al-Nafīsī (Tunis: Dār al-Barrāq, 1990), pp. 129–30.
2. Al-Ghannūshī, Maḥāwir Islāmiyya, op. cit., p. 71.
3. Alan R. Taylor, The Islamic Question in Middle East Politics, op. cit., p. 72.
4. Al-Ghannūshī, Maqālāt, op. cit., p. 91.
5. Raḍwān al-Sayyid, al-Sharī'a wa al-umma wa al-dawla: ishkāliyyāt al-fikr al-islāmī al-mu'āṣir (Malta: Islamic World Studies Centre, 1991), p. 20.
6. Muḥammad Abid al-Jabrī, "al-Ḥaraka al-salafiyya wa-l-jamā'āt al-dīniyya al-mu'āṣira fi-l-maghrib," in al-Ḥarakāt al-islāmiyya al-mu'āṣira fi-l-waṭan al-'arabī, op. cit., p. 234.
7. Leon T. Hadar, 'What Green Peril?' in Foreign Affairs, vol. 72, no. 2, 1993, p. 35.
8. Rāshid al-Ghannūshī, Naẓarāt ḥawla al-dīmūqrāṭiyya fi-l-maghrib al-'arabī: al-insān, vol. 1 (Paris: March 1990), p. 70.

9. Elbaki Hermassi, 'The Islamicist Movement and November 7' in Zartman, Tunisia: The Political Economy of Reform, op. cit., p. 196.

10. Lars Rudebeck, Party and People: A Study of Political Change in Tunisia, (London: C. Hurst and Company, 1967), p.257.

11. See accounts of examples of such arguments between Bourguiba and some of his ministers in the early sixties in Charles A. Micaud's book Tunisia: The Politics of Modernisation (with L. Brown and C.H. Moore), (London: Pall Mall Press, 1964), pp.99–101.

12. Ibid. p. 104.

13. Ibid. pp. 97–8

14. Elbaki Hermassi, 'The Islamicist Movement and November 7' in Zartman, Tunisia: The Political Economy of Reform, op. cit., p. 196.

15. 'Abd Allāh al-Nafīsī, "al-Ikhwān al-muslimūn fi Misr: al-tajriba wa-l-khaṭa'," in al-Ḥaraka al-islāmiyya: al-ru'ya al-mustaqbaliyya, op. cit., p. 217.

16. Arnold H. Green, The Tunisian Ulama 1873–1915, Leiden: E.J. Brill, 1978.

17. Ibid., p. 14.

Appendix 1

The Founding Manifesto of *Ḥarakat al-Ittijāh al-Islāmī* (1981)

Introduction

The Islamic world, of which Tunisia is a part, is currently witnessing the worst forms of distortion (*istilāb*) and alienation regarding its identity and interests. Since the Middle Ages, degenerating influences have been active in eating away at the state of our nation (*umma*), compelling it to relinquish its pioneering and influential mission. At times this has been for the benefit of the colonising West, while at others for internal ruling minorities, separating it [the *umma*] from its roots and contradicting the interests of its people.

Throughout this entire period the primary target has been Islam, which is the centre of our identity as a civilisation and the basis of our collective conscience. It has been gradually and slowly marginalised—at times brutally and publicly—from its position of orienting and actively ruling over our lives. Despite being the decisive factor behind the most brilliant achievements of our civilisation, and the vital role it has played in the struggle to evict the coloniser, Islam today is almost no more than a symbol, beset by cultural, moral and political dangers. This is the result of all it has been subjected to, and of negligence of and attacks on its values, institutions and men of distinction, particularly in more recent times.

In addition to these civilisational facts (which we share with all other countries of the Islamic world), despite gaining independence on paper, Tunisia witnessed particular circumstances in the late fifties and during the sixties and seventies which were marked by crises, an increase in social confrontation, and the suspension of its overall development. This situation was further exacerbated by the unitarian nature of the ruling political movement (the Destour Party), and its continued and gradual domination over the authorities, institutions and popular organisations on the one hand; and also as a result of hasty and changeable economic and social policies dependent on international interests, as opposed to the national interests of our people, on the other.

In such a climate, the Islamic Trend in Tunisia emerged in the early seventies, once all the reasons for its existence were in place and its necessity had become apparent. This movement has contributed to restoring respect for Islam as a way of thought, culture and behaviour, as well as re-establishing respect for the mosque. It has also participated in reactivating cultural and political life, and has breathed new life into it by emphasising our identity, increasing awareness of [our national] interests, and asserting pluralism by making it a reality.

By its actions and numerous declarations, the Islamic Trend has shown its adherence to the nation's identity and representation of its people's hopes and aspirations, thus attracting the support of vast sections of the have-nots, youths and intellectuals. Its rapid growth has been a source of concern to observers, and it has been closely monitored by forces and governments both inside and outside [the country]. Despite its rational and composed efforts to seek out the most effective means of achieving evolution and change, the Trend has been the subject of various unfounded accusations and negative publicity campaigns, orchestrated by the ruling authorities and the official and semi-official media. In an attempt to suppress its voice, these campaigns reached a level of unjustified aggression against the Trend's media, and increased levels of oppression by putting its members on trial, widening police supervision and investigations, and by opening up prisons and detention centres for its young [supporters]—places of beatings, torture and vilification.

The persistence of the reasons behind the declining political, economic and cultural situation in our society vindicates the Islamists' legitimate sense of their divine, national and human responsibility to continue and improve their efforts—out of necessity, for the sake of the country's true liberation and for its development, founded on the just principles of Islam and adherence to its correct path.

Some may interpret these actions as mixing religion with the world of politics, and as a gateway leading to the monopolisation of the Islamic notion (*ṣifa*), thereby denying it to others. This understanding not only is the expression of a foreign ecclesiastical concept alien to our indigenous culture, but also helps to sustain, in the contemporary context, the sense of a loss of direction which our nation has historically undergone.

However, the Islamic Trend Movement does not present itself as the official voice of Islam in Tunisia and does not wish to be assigned that label in the future. While recognising the right of all Tunisians to honest and responsible dealings with religion, the movement believes that it has the right to adopt a comprehensive interpretation of Islam, representative of the theological base from which all those intellectual views and political, economic and social policies defining the identity of this movement and directing its strategic orientations and tactical positions are derived. In this context, the Islamic Trend Movement clearly defines the boundaries which limit its areas of responsibility. It is not, then, accountable for the various actions and positions [taken by other groups] which arise from time to time, other than in what the movement officially adopts, regardless of the religious claims of those behind these actions or their Islamic banners.

To confirm this position on the one hand, and to meet the gravity of the responsibility and the needs of this era on the other, it is obligatory for the Islamists to enter a new era of action and organisation, so as to enable them to bring together their supporters, and to enlighten, educate and use them to serve the causes of their people and the nation. This work must be carried out within a movement that has clear objectives, specified methods, defined institutions and representative leaders.

The Islamic Trend Movement, which was cut off from the broad support of the Muslim peoples on account of oppression and terrorism, hopes that the contribution of its supporters will be more original and comprehensive in the future.

Major Objectives

1. To revive the Islamic personality of Tunisia so that she may resume her traditional role as a major centre of Islamic civilisation and culture and to put an end to progressive alienation of the population and the practice of slavish imitation of the West.
2. To reformulate Islamic thought, taking into account the fundamental principles of Islam, the requirements of man's continuous evolution and changing circumstances.
3. To reassert popular will as a political force and, in so doing, to reject local paternalism and foreign influence.
4. To establish a system of social justice based on the principle that, although everyone should have the right to benefit from his own endeavours, subject to the public interest, everyone should also enjoy the right to receive what he needs—"to each according to his efforts and to each according to his needs"—so that the masses attain their legitimate rights to live in dignity, removed from all kinds of exploitation and submission to international economic powers.
5. To contribute to the revival of the political and civilisational unity of Islam nationally, regionally, within the Arab world, and internationally, thus saving our people and the whole of humanity from psychological alienation, social injustices and international imperialism.

Methods of Action

To achieve the above objectives the movement will undertake the following measures:

- To revive the mosque as a centre for worship and comprehensive popular mobilisation, as was the case in the time of the Prophet, and as an extension of the role that the biggest mosque, Jāmi' al-Zaytūna, used to play in protecting Tunisian identity and reinforcing the status of our country as a leading international centre of civilisation.
- To invigorate intellectual and cultural life by, for example, organising conferences; encouraging writing and publishing; consolidating and elaborating on Islamic principles and values in the domains of literature and culture in general; encouraging scientific research; and supporting responsible, rather than immoral and hypocritical, media.
- To support Arabisation in education and administration, whilst remaining positive towards foreign languages.
- To reject violence as a means of bringing about change, and to base this struggle on the principles of *shūrā* as a mechanism for solving differences in thought, culture and politics.
- To reject the one-party system because of its implications, i.e., its denial of the human being's will, its incapacitation of people's capabilities, and its steering the country down the path of violence. The rejection of the one-party system must be coupled with the recognition of the rights of all popular forces to exercise freedom of expression, organisation and all other legitimate rights. We shall co-operate in this regard with all national forces.
- To crystallise Islamic social principles by devising contemporary practical applications; to analyse the current situation of the Tunisian economy in order to define the problems of injustice and its causes; and to elaborate alternative solutions.
- To side with the *mustaḍ'afīn* [the impoverished] masses of workers, labourers and all deprived people, in their struggle against the *mustakbirīn* and the rich.
- To support the trade union movement in order to safeguard its independence and its capability to achieve national liberation in all its social, political and cultural dimensions.
- To adopt a comprehensive understanding of Islam and a commitment to political action unaffected by secularism and opportunism.
- To liberate the Muslim conscience from [the complex of] civilisational defeat by the West.
- To crystallise and implement the contemporary form of the Islamic political system; to ensure that all national issues are debated objectively

within the theological and historical contexts of the Maghreb, the Arab and Islamic world, and within the context of the world of the *mustaḍ'afīn* in general.

— To strengthen relations of fraternity and co-operation among all Muslims: in Tunisia, the Maghreb and Islamic world as a whole.

— To support and assist liberation movements from around the world.

Tunis, 6 June 1981

Appendix 2

The Manifesto of *Al-Nahḍa* Movement of Tunisia (1988)

Part One: Foundations and Aims

Article 1

Among those Tunisian citizens who have agreed or will agree on this charter and who believe in its goals, which will be stated for an indefinite period, a party has been created under the name *Ḥarakat al-Nahḍa*, which is subject to the law of 3 May 1988 and the following articles mentioned therein.

Article 2: Aims

Al-Nahḍa will endeavour to effectuate the following goals:

(a) The political domain

(i) To consolidate the republican regime and its foundations, to safeguard civil society, to implement the principle of popular sovereignty and to establish *shūrā*.

(ii) To implement freedom, as the basic value embodying the dignity that God has conferred upon the human being, by enhancing civil and personal liberties and human rights, and by promoting the principle of the independence of the judiciary and the neutrality of the administration.

(iii) To operate a foreign policy founded on the country's sovereignty, its unity and independence from interference at all levels, and to operate international relations on the bases of positive non-alignment, mutual respect and the right of the people to self-determination, justice and equality.

(iv) To promote co-operation among Arab and Islamic countries, and to enhance solidarity and unity.

(v) To promote the spirit of Arabic and Islamic unity and awareness of the fundamental issues of the *umma*, so as to put an end to hostilities, divisions and secession; to concentrate all efforts on our most decisive issues; to strive to bring about comprehensive unity and to support all serious steps towards this; and to give supreme importance to the unity of the countries of the Arab Maghreb.

(vi) To strive to liberate Palestine, and to consider this a central mission and duty implicit in the necessity of standing up to colonial Zionist aggression, which has planted an alien entity at the heart of our Arab nation so that it might be an obstacle to its unity, and in reflection of the civilisational struggle between our nation and its enemies.

(vii) To support liberation causes in the Arab-Islamic world and in the world as a whole; to struggle against colonial and segregationist policies in Afghanistan, Eritrea, South Africa and others; to express solidarity with all oppressed people in their fight for liberation, and to oppose all situations founded on injustice and tyranny.

(viii) To work for the development of co-operation with African countries and to ensure a strategic orientation for our country; to work to neutralise the Mediterranean basin from the struggle between the forces of hegemony, and to eliminate all factors of tension in the region; to help consolidate relations of mutual understanding and co-operation between all peoples; and to promote world peace founded on justice.

(b) The economic domain

(ix) To build a strong, integrated national economy, which will rely essentially on our resources, lead to self-sufficiency, and fulfil our basic needs; to ensure a balance among the regions [of the country]; to help effectuate integration and close co-operation in the Maghreb, and with the Arab-Islamic world.

(x) To achieve complementarity and balance among the national sectors, i.e., the public, the private and the co-operative, in such a way that serves the public interest.

(xi) To emphasise that work is the basis of any gain and the condition for progress, and that it is a right and a duty; to work to build an economic system based on human principles; to distribute the country's wealth according to the principle of "to each according to his efforts and to each according to his needs" (i.e., although everyone should have the right to benefit from his own endeavours—subject to the public interest—everyone should also enjoy the right to receive what he needs); and to eradicate all social distinctions based on exploitation, the accumulation of wealth, monopolies and other illegitimate means.

(c) The social domain

(xii) To ensure social services that guarantee the essentials for all—their rights to food, health, education, housing and all other basic requirements for a dignified life—so as to safeguard social cohesion and development, and to free the potential for spirituality, beauty and innovation by combining the guarantee of rights with the execution of duty.

(xiii) To support all popular organisations and protect their unity, existence and the democratisation of their internal decisions; to respect their independence so that they may express the needs of their members and defend their interests; and to assist in offering guarantees or protection against all forms of dictatorship.

(xiv) To safeguard the entity of the family, which is the basis of a healthy society, and to ensure that relations within it are based upon affection, mercy, cohesion and mutual respect; to safeguard the sanctity of marital relations; and to ensure the correct and necessary environment for the proper raising and nurturing of children.

(xv) To raise the status of women and reinforce their positive role on the social, cultural, economic and political scenes, so that they may contribute effectively to the development of society without imitation or degradation, and that they may realise their own identity; and to protect their dignity by rejecting all expressions of immorality.

(xvi) To protect the youth, the very heart of the nation, and to prepare them well for the challenges of progress and development with a sound education; to open up prospects for their effective contribution towards [the country's] overall development; to facilitate their integration into society; to guarantee employment for them; and to encourage and facilitate marriage.

(xvii) To base social relations on the civilisational principles of our countries by promoting moral values, so that this may ensure a spirit of fraternity and mercy within society and protect it from harm.

(d) The cultural domain

(xviii) To assert and reinforce the Arab-Islamic identity as one of the conditions for its revival; and to accord it the status it deserves by implementing the requirements of the country's constitution and laws, in respect of the fact that Islam is both a set of values for civilisation and a way of life, and that Arabic is the language of national culture.

(xix) To ensure the required environment for a comprehensive scientific and intellectual revival, based upon both what is well established among the teachings of Islam and the requirements of changing times, so as to bring an end to the state of backwardness, decadence, dependence and alienation; and to bring about intellectual and cultural activities that will enlighten the mind, refine tastes and behaviour, and enhance Tunisia's civilisational role.

(xx) To adopt the Arabic language in the educational, administrative and cultural domains, and to raise it to the extent that it may be a tool for a civilisational revival that will help in uniting the Islamic *umma*; and to facilitate positive interaction with world cultures without a sense of defeatism, self-deprecation or isolation.

(xxi) To ensure the conditions necessary for the encouragement of scientific research and its promotion; to respect scholars, researchers and inventors, and to accord to them the status they deserve. This is in accordance with our belief in the important role that science can play in developing the country, supporting its independence, and in seeking an intellectual and scientific revival, so as to make the actions of the human being compatible with the laws of nature and the course of history.

(xxii) To implement a media policy based on respect for freedom of thought and expression; to promote the spirit of originality and innovation; and to guarantee the conditions necessary for independent, honest and responsible media that contribute to the progress of the country and foster its identity.

(xxiii) To encourage literature, the arts, and the practising of sports, so that they may fulfil their role in promoting morality and safeguarding fitness, the advancement of spirituality, and supporting the basis of revival.

Tunis, June 1988

Appendix 3

The Islamic Basis of Our Foreign Policy

Extract from an internal *Nahḍa* memo on foreign relations

The foreign policy of an Islamic group on its relations with other groups, states and individuals is governed, like all other relations, by the principles of the Islamic *sharī'a*. Believers must always ascertain God's ruling on any issue before taking action. The relevant *sharī'a* principles are as follows:

1. Allegiance and friendship: A Muslim's allegiance must be to God alone, that is to say, to [the principles and the rulings outlined in] His book [the Qur'an], and to His Prophet (the *sunna*), and to fellow Muslims. "Your ally is God, His messenger and the believers" (*al-Mā'ida*: 55). In cases where a Muslim is permitted to lend support to a non-Muslim or form an alliance with one, this must not under any circumstances be done against [the interests of] another believer. "Believers, do not take the unbelievers as allies instead of the believers" (*al-Nisā'*:144) and "Whoever does that shall have nothing to do with God" (*Āl-'Imrān*: 28).
2. Other than the above, there are no restrictions on establishing relations of co-operation or agreements between Muslims and non-Muslims, as long as the latter are not in a state of open war against Islam or the Muslims. The Qur'an says:

> God does not forbid you to be kind and equitable to those who have neither made war on your religion nor driven you from your homes. God loves the equitable. But He forbids you to show friendship towards those who have fought against you on account of your religion and driven you out of your homes and abetted others to do so. Those who make friends with them are transgressors (*al-Mumtaḥina*: 8–9).

God also says in the Qur'an: "You shall not find any people who believe in God and the Last Day who would be on friendly terms with those who oppose God and His messenger, even if they were their fathers, their sons, their brothers or their kindred. God has inscribed faith in their hearts, and strengthened them with a spirit of His own"(*al-Mujādala*: 22).

3. The blood and possessions of those who are not in a state of war with the Muslims or against whom we have declared no war, are safe with us (see: *al-Mumtaḥina*: 8).
4. All Muslims are one nation: "This nation of yours is one nation, and I am your Lord, so worship Me"(*al-Anbiyā'*: 92). Our duty, therefore, is to support the believers and work for their unity, solidarity and integrity as much as possible. In principle, all believers should form one unit in times of peace or war.
5. Muslims today are not united under one leader. Rather, they operate in groups or as individuals. As a consequence, they cannot give any unified commitments; rather, they may be diverse. The principles governing the relations of Muslims should be in all circumstances based on the saying, "A Muslim is the brother of a Muslim; he does not oppress him, nor does he fail him, nor does he betray him"; and on the principle "There shall be no harming [of one man by another] in the first instance, nor in requital [in Islam]."

   If the interests of a Muslim group require that they establish relations with a non-Islamic régime which is persecuting [another] Islamic group, then is such a relationship permissible? It is preferable not to associate with such a régime. However, if it is unavoidable, in order to protect the lives and honour of other Muslim groups, then it is permissible, providing it will not be at the expense of other Muslims, causing them to be let down or aggression to be committed against them. For a Muslim should not fail another Muslim; rather, he should protect him, he should try to intervene to deter his enemy, and he should strive to rescue him. He must never, under any circumstances, side against him.
6. The status of a Muslim living in a non-Muslim country is governed by a treaty in the form of the entry visa he is granted. This represents a statement of mutual agreement between two sides, which binds him to respect the law of the land in return for receiving a promise of protection from the state.

Tunis, 1988

Appendix 4

Dedication

From Ghannūshī's *Civil Liberties in the Islamic State* (1993)

I dedicate this book to all those who, after Allah, have favoured myself, the Islamic nation and all humanity, by serving Islam and the causes of rightness and justice world wide, whether listed here or not. I dedicate it even to those I do not know, in recognition of their deserving praise for their efforts.

I dedicate it to the spirit of my beloved mother, Zayna al-Zarībī, and to my beloved father, Imḥammad al-Ghannūshī; to the spirit of my dear brother, Professor Mukhtār al-Ghannūshī; to my beloved wife Fāṭima, and to my beautiful children, Tasnīm, Mu'ādh, Sumayya, Yusrā, al-Barā' and Intiṣār. They are Allah's gift to me, and have embraced me with love and warmth and spurred me on in the fight.

I dedicate it to the city of Damascus which, with the help of an unknown soldier, the Damascene pharmacist Brother Muḥammad Amīn al-Mujtahid, witnessed my second birth; also to all my teachers at Damascus University, and to my other teachers both in my country and abroad, among whom I give special mention to my late master al-Shaykh Muḥammad Ṣaliḥ Al-Nayfar, my master al-Shaykh 'Abd al-Qādir Salāma, and my master al-Shaykh Muḥammad al-Ikhwa.

Also to my spiritual fathers, led by the martyr Ḥasan al-Bannā, Mawlānū Abu al-A'lā al-Mawdūdī, the martyr Sayyid Quṭb, and our teacher Mālik ibn Nabī; to the innovator and leader al-Shaykh Ḥasan al-Turābī, and to the leader of the contemporary Islamic revolution al-Imām al-Khumaynī; to the renowned scholar [Muḥammad Bāqir] al-Ṣadr; to the martyr 'Alī Sharī'atī; to the poet of Islam Muḥammad Iqbāl, and to the Shaykh, renowned scholar and symbol of the Islamic *da'wa* in North Africa ['Abd al-Hamīd] ibn Badīs; to al-Shaykh 'Uthmān Fūdī; to the Mahdī of Sudan; to 'Abd al-Qādir al-Jazā'irī; and to 'Abd al-Karīm al-Khaṭābī and to 'Umar al-Mukhtār, all symbols of the Islamic *jihād* against colonialism in Africa; to the *mujāhid* and intellectual, the Islamic President Ali Izebegovitch; and to Professor Roger Garaudy, among the vanguard of Western Islam; and to Khayr al-Dīn al-Tūnisī, the forerunner of Islamic modernisation.

To the leaders of the Islamic Alliance, led by Jamāl al-Dīn al-Afghānī, Muḥammad 'Abduh, Rashīd Riḍā, Shakīb Arsālan, al-Shaykh ['Abd al-'Azīz] al-Tha'ālibī, and al-Shaykh [Muḥī al-Dīn] al-Qalībī; to the renowned scholar Muḥammad al-Ṭāhir ibn 'Āshūr; and to the symbol of God's unity, al-Shaykh Muḥammad ibn 'Abd al-Wahhāb.

Also to the martyrs of Islam everywhere, led by the dear Brother 'Abd Allāh Azzām, martyr of Islam in the *jihād* against communist colonialism; the renowned scholar Ismā'īl al-Fārūqī, martyr of Islam in the West; to Malcolm X, martyr of Islam in the *jihād* against racial segregation; to the Bosnian martyrs, and others who have paved the future path of Islam with their pure bodies, including the Islamic martyrs in North Africa, beginning with the great conqueror 'Uqba ibn Nāfi; to the martyrs of the National Movement, led by Farḥāt Ḥashād, followed by the martyrs of the second war of liberation, led by Muṣṭafā Abu Ya'lā; to 'Uthmān ibn Maḥmūd, 'Abd al-Ra'ūf al-'Arībī and Muḥammad al-Manṣūrī; and all who have followed the path of martyrdom and light.

And to the Islamic prisoners in Tunisia, led by Dr al-Ṣādiq Shūrū, 'Alī al-'Arīdh, al-Ḥabīb al-Lūz, Dr Ziyād al-Dūlātlī, Ḥammādī al-Jabālī and all current and former victims of oppression.

And to Islamic prisoners all over the world, led by the *mujāhids* al-Shaykh 'Abbāsī Madanī and 'Alī Belḥāj, to the pious al-Shaykh 'Abd al-Salām Yāsīn, to the *mujāhid* prisoner Ghulām A'ẓam, *amīr* of *al-Jamā'a al-Islāmīyya* in Bangladesh; and to al-Shaykh Aḥmad Yāsīn in the land of *isrā'* and *mi'rāj*, and his Muslim Brothers who were exiled to Lebanon, all symbols of heroism, resistance, and martyrdom, exposing the values of hypocrisy, injustice and ignorance dominating the world.

To the leaders of the Islamic women's liberation movement, led by Shaykha Zaynab al-Ghazzālī in Egypt and Shaykha Su'ād al-Fātiḥ in Sudan; to the young women of Bosnia, through whom the whole Islamic nation and every honest, free man was raped; to their Tunisian Sister Zahra al-Tīs, martyr of the student movement; and to the mother of our Sister Jalīla al-Khumsī, steadfast behind bars after her leg was broken under torture; all of them witnesses to the oppressive nature and moral bankruptcy of the new world order.

To the prisoners of conscience and freedom of all faiths, and in every place. To the fighters against local and international tyranny, led by Mr Gonzalo, leader of the Shining Path movement, fighter against the symbols of the military junta in Peru.

To those organisations, personalities and forces fighting for justice, peace, dialogue and co-operation among peoples and civilisations, especially the Tunisian League for the Defence of Human Rights and Amnesty International; to those in the vanguard of Western thought, fighting for a better understanding and for better dealings with Islam, its preachers and its nation. From among them I especially mention the French professor François Burgat, the American professor John Esposito, the Englishman Ernest Gellner and the German Mr Hoffman.

To those holding the embers of Islam, to those carrying its light everywhere, in dignity and moderation, reacting to humankind's concern for a world that will be purified from *kufr*, poverty, oppression, injustice and tyranny.

To every person who has taught me a letter, or guided me to write, or deterred me from doing wrong, or who has offered any kind of help in the catastrophes enforced on the *umma* and all struggling nations, if only in the form of a pleasant word.

To all those mentioned and to all of those like them I dedicate my book, in acknowledgement of their favours towards me—after Allah's. Also, that it may be a fighting tool in their hands; that it may assert the relationship of kinship; that it may facilitate meetings, dialogue, love and co-operation among them and those following their path—for the sake of humanity, which deserves to receive this dignified and divine call:

> O Mankind! We created you from a single [pair] of a male and a female, and made you into nations and tribes that you may know each other [Not that you may despise each other]. Verily the most honoured of you in the sight of God is [he who is] the most righteous of you (*al-Ḥujurāt*: 13).

Bibliography

Works in English

Anderson, Lisa, *The State and Social Transformation in Tunisia and Libya 1830–1980*, Princeton Studies on the Near East (Princeton, NJ: Princeton University Press, 1986).

Burgat, François and William Dowell, *The Islamic Movement in North Africa* (Austin: University of Texas Press, 1993).

Burrell, R. M., ed., *Islamic Fundamentalism* (London: Royal Asiatic Society, 1989).

Dessouki, Ali E. Hillal, ed., *Islamic Resurgence in the Arab World* (New York: Praeger, 1982).

Dīn, Khayr al-, *The Surest Path* translated by Leon Carl Brown (Massachusetts: Harvard University Press, 1967).

Donahue, J. and Esposito J., eds., *Islam in Transition: Muslim Perspectives* (New York: Oxford University Press, 1982).

Esposito, John, *The Islamic Threat: Myth or Reality?* (New York: Oxford University Press, 1992).

———, ed., *Voices of Resurgent Islam* (New York: Oxford University Press, 1983).

———, *Islam and Politics* (New York: Oxford University Press, 1987).

Ferdinand, Klaus and Mehdi Mozaffari, *Islam: State and Society* (London: Curzon Press, 1988).

Green, Arnold H., *The Tunisian Ulama 1873–1915*, (Leiden: E. J. Brill, 1978).

Hourani, Albert, *Arabic Thought in the Liberal Age* (Cambridge: Cambridge University Press, 1983).

Hunter, Shireen T., ed., *The Politics of Islamic Revivalism: Diversity and Unity* (Bloomington: Indiana University Press, 1988).

Kemal, Mustafa, *A Speech* (Leipzig: K.F. Koehler, 1929).

Lewis, Bernard, *The Political Language of Islam* (Chicago and London: University of Chicago Press, 1988).

———, *The Middle East and the West* (London: Weidenfeld and Nicolson, 1963).

Mawdūdī, Abu al-A'lā al-, *Islam and Modern Civilisation* (Cairo: Dār al-Anṣār, 1978).

Mehran, Tamadonfar, *The Islamic Polity and Political Leadership: Fundamentalism, Sectarianism and Pragmatism* (Boulder: Westview Press, 1989).

Micoud, Charles A., with Leon C. Brown and Clemont H. Moore, *Tunisia, The Politics of Modernisation* (London: Pall Mall Press, 1964).

Parker, Richard B., *North Africa: Regional Tensions and Strategic Concerns*, (New York: Praeger Publishers, 1987).

Piscatori, James, ed., *Islam in the Political Process* (Cambridge: Cambridge University Press, 1983).

Rudebeck, Lars, *Party and People: A Study of Political Changes in Tunisia*, (London: C. Hurst and Company, 1967).

Said, Edward W., *Covering Islam: How the Media and Experts Determine How We See the Rest of the World* (New York: Pantheon Books, 1981).

Sharabi, Hisham, *Arab Intellectuals and the West: The Formative Years 1875–1914* (Baltimore: Johns Hopkins University Press, 1970).

Tamimi, ed., *Power-Sharing in Islam* (London: Liberty Publications, 1993).

Taylor, Alan R., *The Islamic Question in Middle East Politics* (Boulder: Westview Press, 1988).

Voll, J., *Islam: Continuity and Change in the Modern World* (Boulder: Westview Press, 1982).

Zartman, I. William, ed., *Tunisia: The Political Economy of Reform*, SAIS African Studies Library (Boulder and London: Lynne Rienner Publishers, 1991).

No author, 'The National Pact' (Tunis: Tunisian External Communication Agency, 1992).

Works in Arabic

Arkūn, Muḥammad, *al-Fikr al-islami: naqd wa ijtihād* (London: Dār al-Sāqī, 1990).

Ash'ari,'Alī ibn Ismā'īl al-, *Maqālāt al-islāmiyyīn* (Cairo: Maktabat al-Nahḍa, 1969).

al-'Awwā, Muḥammad Salīm, *Fi al-niẓām al-siyāsī li al-dawla al-islāmiyya* (Cairo: Dār al-Shurūq, 1989).

Bannā, Ḥasan al-, *Majmū'at rasā'il al-imām al-shahīd Ḥasan al-Bannā* (Cairo: Kutub al-Da'wa, n.d.).

―――, *Islāmunā* (Cairo: Dar al-I'tiṣām, 1977).

―――, *Risāla ilā al-shabāb* (Cairo: Al-Zahra for Arab Media, 1989).

Darwīsh, Qusayy Ṣliḥ al-, *Yaḥduthu fī Tūnis* (Paris: Q. Darwīsh, 1987).

Ghannūshī, Rāshid al-, *Maqālāt* (Paris: Dār al-Karawān, 1984).

―――, *Ḥarakat al-ittijāh al-islāmī fī Tūnis* (Safat: Dār al-Qalam Kuwait, 1989).

―――, *Maḥāwir islāmiyya* (Khartoum: Bayt al-Ma'rifa, 1989).

―――, *al-Ḥurriyyāt al-'āmma fī al-dawla al-islāmiyya* (Beirut: Markaz dirāsāt al-waḥda al-'arabiyya, 1993).

―――, *Min al-fikr al-islāmi fī Tūnis* (Kuwait: Dār al-Qalam, 1992).

―――, *Ḥuqūq al-muwāṭana: ḥuqūq ghayr al-muslim fī al-mujtama' al-islāmī*, 2nd edn (Virginia: International Institute of Islamic Thought, 1993).

―――, *al-Mar'a al-muslima fī Tūnis* (Kuwait: Dār al-Qalam, 1988).

―――et al., *al-Hiwar al-qawmī al-dīnī* (Proceedings of a conference) (Beirut:, Markaz dirāsāt al-waḥda al-'arabiyya, 1989).

Ḥāmidī, Muḥammad al-Hāshimī al-, *Ashwāq al-ḥurriyya* (Kuwait: Dār al-Qalam Kuwait, 1989).

Hammāmī, Ḥamma al-, *Ḍidda al-ẓalāmiyya: fī al-radd 'alā al-ittijāh al-islāmī*, 2nd edn (Tunis: Dār al-nashr li'l-maghrib al-'arabī, 1986).

Harmāsī, Muḥammad 'Abd al-Bāqī al-, *al-Ḥarakāt al-islāmiyya al-mu'āṣira fī al-waṭan al-'arabī* (Beirut: Markaz dirāsāt al-wiḥda al-'arabiyya, 1987).

Ḥawwā, Sa'īd, *al-Madkhal ilā da'wat al-ikhwān al-muslimīn* (N.p.: n.p., n.d.).
'Imāmī, 'Abd Allāh, *Taẓīmāt al-irhāb fī al-'ālam al-islāmī: unmūdhaj al-Nahḍa: al-nash'a, al-taẓīr, al-haykala, al-irhāb* ... (Tunis: Al-Dār al-Tūnisīyya li'l-nashr, 1992).
Jarbī, Jallūl al-, *al-Huwwiyya fī Tūnis al-'ahd al-jadīd* (Tunis: Al-Wikāla al-Tūnisiyya li-l-ittiṣāl al-khārijī, 1992).
Manṣūrī, Walīd al-, *al-Ittijāh al-islāmī wa Burqayba: muḥākamat man li-man?* (Tunis: n.p., 1988).
Mzali, Muḥammad, *Dirāsāt* (Tunis: Tunisian Company for Distribution, 1984).
Nadwī, Abu al-Ḥasan 'Alī al-, *Mādha khasira al-'ālam bi inḥiṭāt al-muslimīn?* (Kuwait: IIFSO, 1978).
Nafīsī, 'Abd Allah al-, ed., *al-Ḥaraka al-islāmiyya: al-ru'ya al-mustaqbaliyya* (Tunis: Dār al-Barrāq, 1990).
Najjār, 'Abd al-Majīd al-, *Ṣirā' al-huwiyya fī Tūnis* (Paris: Dār al-Amān, 1988).
Quṭb, Sayyid, *Fī zilāl al-Qur'ān*, vol. 5 (Beirut: Dār al-Shurūq, 1985).
———, *al-Mustaqbal li hadha al-din* (Kuwait: IIFSO, 1978).
Sayyid, Raḍwān al-, *al-Sharī'a wa al-umma wa al-dawla: ishkāliyyāt al-fikr al-islāmī al-mu'āṣir* (Malta: Islamic World Studies Centre, 1991).
Shābbī, Anas al-, *al-Taṭarruf al-dīnī fī Tūnis* (Tunis: La Presse, 1991).
Tūnisī, Khayr al-Din al-, *Muqaddimat aqwam al-masālik fī ma'rifat aḥwāl al-mamālik* (Beirut: Dār al-Ṭalī'a, 1978).
Van Krieken, *Khayr al-Dīn wa Tūnis:1850–1881* (Tunis: Dar Sahnūn and E.J. Brill, 1988).
Wannās, al-Munṣif, *al-Dawla wa al-mas'ala al-thaqāfiyya fī Tūnis* (Tunis: Dār al-Mīthāq, 1988).
al-Zghal, 'Abd al-Qādir et al., *al-Dīn fī al-mujtama' al-'arabī* (Beirut: Markaz dirāsāt al-waḥda al-'arabiyya, 1990).
Various authors, 'November 7: al-Thawra al-hādi'a' (Tunis: A. Ben Abdallah for Publishing and Distribution, 1992).

Works in French

Balta, Paul, *L'islam dans le monde: dossier établi et présenté par Paul Balta*, 2nd edn, Collection la mémoire du monde (Paris: Editions Le Monde, 1991).
Balta, Paul, with Claudine Rulleau, *Le Grand Maghreb: Des indépendances à l'an 2000* (Alger: Editions Laphomic, 1990).
Bessis, Sophie and Souhayr Belhassen, *Bourguiba: Un si long règne (1957–1989)*, vol. 2 (Paris: JAPRESS, 1989).
Boularès, Habib, *L'islam: la peur et l'espérance* (Paris: Editions J-C Lattès, 1983).
Bourguiba, Habib, *Discours: Tome VII, Année 1959–60* (Tunis: Publications de Secretariat d'Etat à l'Information, 1976).
———, *La Tunisie et la France* (Paris: Julliard, 1954).
Burgat, François, *L'islamisme au Maghreb: la voix du Sud (Tunisie, Algérie, Libye, Maroc)* (Paris: Karthala, 1988).
Djait, Hishem, *La personalité et le devenir Arabo-Islamique* (Paris: Seuil, 1974).
Ganari, Ali al-, *Bourguiba le combattant suprême* (Paris: Plon, 1985).
Garaudy, Roger, *Intégrismes* (Paris: Pierre Belfond, 1990).

Lamchichi, Abderrahim, *Islam et contestation au Maghreb* (Paris: L'Harmattan, 1989).
Mzali, Mohammed, *Tunisie: Quel avenir?* (Paris: Publisud, 1991).
———, *Lettre ouverte à Bourguiba* (Paris: Editions Alain Moreau, 1987).
Sivers, Von, *Islam et politique au Maghreb* (Paris: CNRS, 1981).
Toumi, Mohsen, *La Tunisie de Bourguiba à Ben Ali* (Paris: PUF, 1989).
Touzani, Baccar, *Maghreb et Francophony* (Paris: Economica, 1988).

Journals and Periodicals

Abwāb (Beirut: Summer 1994).
Al-Insān, vol. 1 (Paris: March 1990).
Al-Wiḥda, no. 96 (Rabat: September 1992).
Foreign Affairs, vol. 72, no. 2 (New York: 1993).
Islamiyyāt al-Ma'rifa, vol. 1, no. 2 (Malaysia: September 1995).
Maghreb Review, vol. 7, nos. 1 and 2 (London: January–April 1982).
Middle East Journal, no. 40 (Washington: 1986).
Qirā'āt Siyāsiyya, no. 3–4, (Florida: Summer–Winter 1993).
Revue Parlementaire, no. 928 (March–April 1987).
The Middle East and North Africa 1980–1981, no. 27 (London: The Times, 1980).
Tunisia News, no. 129 (Tunis: 29 May 1995).

Reference Books

Ali, A. Yusuf, *The Holy Qur'an: Translation and Commentary* (Maryland: Amana Corp, 1983).
Annuaire de l'Afrique du Nord, 1979 (Paris: Editions CNRS, 1981).
Encyclopaedia of Islam (Leiden: Leiden University Press, 1971).

Unpublished Documents and Pamphlets

Al-Ghannūshī, Rāshid, "Ilā 'ulamā' al-umma wa jamāhīrihā" (London, 1991).
———, "al-Murāsala," no. 8. (London: 1 December 1995).
Al-Nahḍa, "The Movement of Islamic Tendency in Tunisia: The Facts" (Tunis: n.p., September 1987).
Al-Nahḍa, "al-Ḥadath al-Maghribī" (23 May 1991).
Al-Nahḍa, "Ḥaqā'iq ḥawla ḥarakat al-ittijāh al-islāmi," (Tunis: n.p., 1983).
Ben Ali, Zin al-Abidine, "Declaration of November 7th, 1987."
Ḥizb al-Taḥrīr, "Principles of Ḥizb al-Taḥrīr," 4th edn (N.p.: n.p., n.d.).
Mashhūr, Mus afā, *Tasā'ulāt 'ala al-tārīkh* (n.p., 1983).
"Realities about the MTI" (Tunis: n.p., 1983).

Magazines and Newspapers

Arabia, London, April 1985.
Al-Da'wa, Islamabad, 16 December 1993, and no. 38 (1994).

Al-Fikr, Tunis, no. 16, 7 April 1971.
Al-Ḥayāt, London, 3 August 1994.
Al-Hurriyya, Tunis, 29 September 1991, 9 June 1992.
Al-Majalla, London, 3 October 1990, 13 May 1992.
Al-Mawqif, Tunis, 17 March 1988.
Al-Mujtama', Kuwait, 15 January 1985.
Al-Mustaqbal, Tunis, 26 February 1988.
Al-Rāyya, Rabat, 1995.
Al-Ṣabāḥ, Tunis, 18 March 1988, 27 May 1991.
Al-Sada, Tunis, 3 April 1988
Al-Sha'b, Egypt, 18 September 1990.
Jeune Afrique, Paris, 27 March 1991, 12 June 1991.
Le Figaro, Paris, 28 August 1987, 29 September 1987, 2 August 1994.
Le Maghreb, Tunis, 13 October 1989, 10 November 1989, 7 December 1990.
Le Monde, Paris, 28 March 1987, 6 August 1987, 11 September 1987, 16 September 1987, 29 September 1987.
Le Point, Paris, 21 September 1987.
New York Times, New York, 25 June 1987.
Réalités, Tunis, 1 December 1989, 26 June 1990, 26 October 1990, 12 April 1991, 17 May 1991, 24 May 1991, 10 July 1992.

Interviews

Al-Ghannūshī, Rāshid, personal interviews, 7 November 1992, 15 November 1992, and 23 May 1993.

Index

9 780813 338880

1108320R0

Printed in Great Britain by
Amazon.co.uk, Ltd.,
Marston Gate.